SOLUTIONS MANUAL

CORPORATE FINANCIAL MANAGEMENT

SOLUTIONS MANUAL

CORPORATE FINANCIAL MANAGEMENT

DOUGLAS R. EMERY • JOHN D. FINNERTY

IN CONJUNCTION WITH

ROBERT M. PAVLIK

Southwest Texas State

Prentice Hall, Upper Saddle River, New Jersey 07458

Acquisitions editor: Paul Donnelly
Associate editor: Gladys Soto
Project editor: Richard Bretan
Manufacturing buyer: Arnold Vila

A Pearson Education Company
Upper Saddle River, NJ 07458

Printed in the United States of America

10

ISBN 0-13-723560-7

Prentice-Hall International (UK) Limited,London
Prentice-Hall of Australia Pty. Limited, Sydney
Prentice-Hall Canada Inc., Toronto
Prentice-Hall Hispanoamericana, S.A., Mexico
Prentice-Hall of India Private Limited, New Delhi
Prentice-Hall of Japan, Inc., Tokyo
Pearson Education Asia Pte. Ltd., Singapore
Editora Prentice-Hall do Brasil, Ltda., Rio de Janeiro

CONTENTS

CHAPTER 2 -- ACCOUNTING, CASH FLOWS, AND TAXES

A1. a. An **income statement** reports the profit (loss) of a firm during a specific period of time.

b. A **balance sheet** reports the financial condition of a firm at one point in time.

c. A **statement of cash flows** shows how the cash position of the firm has changed during the period covered by the income statement.

A2. The notes to a firm's financial statements form an integral part of the financial statements because they disclose the significant accounting policies used to prepare the financial statements and provide additional detail concerning several of the items in the financial statements.

A3. The basic balance sheet identity is: Assets = Liabilities + Stockholders' Equity.

A4. a. Current assets are assets that mature or are expected to be realized as cash within 1 year.

b. Current liabilities are liabilities that come due or are expected to be paid out in cash within 1 year.

A5. **Cash flow** is the sum of net income before extraordinary items and noncash expenses. Earnings represent an accounting measure of profit and include the effect of noncash items such as depreciation, deferred income taxes, minority interest, etc., while cash flow includes only those items that affect the cash position of the firm.

A6. **Working capital** is the current assets of the firm. It is typically used to mean *net working capital* which is the difference between current assets and current liabilities. Working capital measures the firm's liquidity, that is, its ability to meet its short-term obligations as they come due.

A7. Four factors that affect the likelihood that the market and book values of an asset will differ are the following: (1) the time since the asset was acquired (2) inflation (3) the asset's liquidity, and (4) whether the asset is tangible or intangible. At the time of its acquisition, an asset's initial book value is likely to be similar to the market value. Over time, market value can diverge significantly from book value since changes in book value are specified by GAAP rather than by economic considerations. Generally, the more time that has passed since an asset was acquired, the greater the chance the asset's market value will differ from its book value. When prices change as a result of inflation, market values of existing assets also change to reflect the difference in purchasing power. The greater the rate of inflation, the greater the likelihood that book values will differ from market values. Liquidity affects market values because, on average, less liquid assets have higher transaction costs if they are sold. This results in greater uncertainty about the net proceeds from their sale. Therefore, if all else is equal, the current market values of less liquid assets can differ more from their book values. Lastly, intangible assets tend to be unique, and there are not active markets for selling them. As a result, the current market value of an intangible asset is especially likely to differ from its book value.

A8. Two factors that affect the likelihood that the market and book values of a liability will differ are (1) remaining maturity of the liability and (2) the financial health of the firm. In a healthy firm, the market value of a liability will essentially equal its book value when it comes due. In contrast, the market value of liabilities that do not have to be repaid for a long time will reflect current economic conditions as well as expectations about the future. The market values of liabilities of financially distressed firms are likely to be below their book values, reflecting the possibility that the distressed firm may not be able to meet its obligations.

A9. Economic income is the total return on an investment, made up of any cash inflow plus the change in the value of the assets and liabilities. Net income from the income statement can differ significantly from economic income, because net income is not a cash flow and because GAAP changes in the firm's assets and liabilities do not reflect changes in market values.

B1.

Rimbey Sporting Goods
Balance Sheet
(Thousands of Dollars)
January 31

ASSETS		TOTAL LIABILITIES & STOCKHOLDERS EQUITY	
Current Assets		Current Liabilities	
Cash and equivalents	$ 200	Accounts payable	$ 500
Accounts receivable	600	Notes payable	300
Inventories	700	Other current liabilities	500
Total current assets	1500	Total current liabilities	1300
Fixed assets		Long-term liabilities	
Net plant and equipment	2500	Long-term debt	1100
Total assets	$4000	Total liabilities	$2400
		Stockholders' equity	
		Common stock	100
		Retained earnings	1500
		Total liabilities and stockholders' equity	$4000

Rimbey Sporting Goods
Annual Income Statement
(Thousands of Dollars)

Sales	$ 6000
Costs of goods sold	3700
Gross profit	2300
Selling, general, and administrative expenses	1200
Depreciation	300
Earning before interest and taxes	800
Interest expense	100
Taxes	300
Net income	400
Dividends on common stock	150
Addition to retained earnings	$ 250

B2.

Rimbey Sporting Goods
Statement of Cash Flows
(Thousands of Dollars)

CASH FLOWS FROM OPERATING ACTIVITIES	
Net income	$ 400
Depreciation and amortization	300
Accounts receivables decrease (increase)	(100)
Inventories decrease (increase)	(150)
Accounts payable increase (decrease)	50
Net cash provided by (used in) operating activities	500
CASH FLOWS FROM INVESTING ACTIVITIES	
Purchase of plant and equipment	(500)
Net cash provided by (used in) investing activities	(500)
CASH FLOWS FROM FINANCING ACTIVITIES	
Notes payable increase	50
Issuance of long-term debt, net	200
Increase in other long-term liabilities	(50)
Cash dividends (common stock)	(100)
Net cash provided by (used in) financing activities	100
Net increase (decrease) in cash and equivalents	100
Cash and equivalents at beginning of year	100
Cash and equivalents at end of year	$ 200

B3. Deducting the extraordinary gain from net income and adding back the loss from discontinued operations gives the net income before extraordinary items and income (loss) from discontinued operations, which is $130 million (= 120 - 15 + 25). Then, to find the EBIT, this amount is adjusted by adding interest expense of $20 million. Thus, EBIT = $150 million (= 130 + 20).

B4.

Dominion Resources, Inc.
Annual Income Statement
(Thousands of Dollars)
Year Ended December 31, 1996

Sales revenue	$ 700
Costs of goods sold	485
Selling expenses	30
Administrative expenses	125
Depreciation expense	100
Interest expense	20
Earnings before taxes	(60)
Taxes @ 40%	0
Net income	(60)
Dividends on common and preferred stock	15
Addition to (reduction in) retained earnings	$(75)

B5. a. Gross Profit = Net Sales - Cost of Goods Sold = 86,656 - 39,268 = $47,388 million

Operating Income = Gross Profit - Operating Expense = 47,388 - (10,535 + 12,272 + 16,617)
= $7964 million

Net Income before Extraordinary Items = Net Income +(-) Extraordinary Losses (Gains)
= 3510 - 535 + 2545 = $5520 million

B6. a. Market Value of Assets = 1200 + (3 × 3000) = $10,200

b. Market Value of Liabilities = 800 + (0.9 × 1000) = $1700

c. Market Value of Equity = 10,200 - 1700 = $8500

B7. a. Working capital = 1200 - 800 = $400

b. Current ratio = 1200/800 = 1.5x

B8. a. Taxable income = 15,000,000 - 6,000,000 - 2,500,000 - 500,000 -200,000
+ 0.3×(150,000) + 400,000
= 6,245,000

Tax liability = 22,250 + 0.34 ×(6,245,000 - 335,000)
= $ 2,031,650

b. Marginal tax rate = 34%

Average tax rate = 2,031,650/6,245,000 = 32.5%

B9. a. Tax liability = 3412.50 + 0.28 ×(25,000 - 22,750) = $4042.50

Marginal tax rate = 28%

Average tax rate = 4042.50/25,000 = 16.17%

b. Tax liability = 12,470.50 + 0.31 ×(75,000 - 55,100) = $18,639.50

Marginal tax rate = 31%

Average tax rate = 18,639.50/75,000 = 24.85%

c. Tax liability = 31,039.50 + 0.36 ×(125,000 - 115,000) = $34,639.50

Marginal tax rate = 36%

Average tax rate = 34,639.50/125,000 = 27.71%

B10. a. Tax liability = 22,250 + 0.39 ×(250,000 - 100,000) = $80,750

b. Marginal tax rate = 39%

Average tax rate = 80,750/250,000 = 32.30%

c. Additional taxes = 0.39×(10,000) = $3900

d. Additional taxes = 0.39×(0.3)×(10,000) = $1170

B11. a. Taxable income = 40,000 - 4900 - 6350 = $28,750

Ted and Alice's tax liability = 3412.50 + 0.28×(28,750 - 22,750) = $5092.50

b. Taxable income = 20,000 - 2450 - 3800 = $13,750

Ted's tax liability = 0.15×(13,750) = $2065.50

Alice's tax liability = $2065.50

Ted and Alice's total tax liability = $4125.00

c. No impact on this tax obligation; interest on munis is tax exempt.

d. Tax obligation increase = 0.28×(1000) = $280.00

B.12 Taxable income = 5,000,000 - 2,000,000 - 500,000 -500,000 - 25,000 + 0.3 ×(400,000)
- 200,000 = $1,895,000

Tax liability = 113,900 + 0.34 ×(1,895,000 - 335,000) = $644,300

Marginal tax rate = 34%

Average tax rate = 644,300/1,895,000 = 34%

B13. Working capital measures the difference between current assets and current liabilities, while the current ratio expresses current assets as a percentage of current liabilities. Both are measures of liquidity, and as long as there is a positive amount of working capital, the current ratio will exceed 1.0.

C1. Economic income = 450,000 + 300,000 = $750,000

Economic income differs from accounting net income because net income is not a cash flow, and GAAP changes in the firm's assets and liabilities do not reflect changes in market values.

Appendix Exercise

Liquidity Ratios

	1996	1995
Current ratio =	376.6/162.7 = 2.31x	334.1/108.6 = 3.08x
Quick ratio =	(376.6 - 133.9)/162.7 = 1.49x	(334.1 - 118.8)/108.6 = 1.98x
Working capital ratio =	(376.6 - 162.7)/546.9 = 39.1%	(334.1 - 108.6)/485.8 = 46.4%
Cash ratio =	9.5/580.4 = 1.6%	12/501.1 = 2.4%

Asset Activity Ratios

	1996	1995
Receivables turnover =	546.9/233.2 = 2.35x	485.8/203.3 = 2.39x
Days' sales outstanding =	365/2.35 = 155.3 days	365/2.39 = 152.7 days
Inventory turnover =	286.3/133.9 = 2.14x	247.3/118.8 = 2.08x
Days' sales in inventory =	365/2.14 = 170.6 days	365/2.08 = 175.5 days
Fixed asset turnover =	546.9/203.8 = 2.68x	485.8/167.0 = 2.91x
Total asset turnover =	546.9/580.4 = 0.94x	485.8/501.1 = 0.97x

Leverage Ratios

	1996	1995
Debt ratio =	(580.4 - 313.7)/580.4 = 0.46x	(501.1 - 293.3)/501.1 = 0.41x
Debt/equity ratio =	266.7/313.7 = 0.85	207.8/293.3 = 0.71
Equity multiplier =	580.4/313.7 = 1.85x	501.01/293.3 = 1.71x

Coverage Ratios

	1996	1995
Times-interest-earned ratio =	51.7/7.7 = 6.71x	37.9/8.0 = 4.74x
Fixed charge coverage ratio =	(51.7 + 0)/7.7 = 6.72x	37.9/8.0 = 4.74x

Note: A tax rate is required to compute the cash flow coverage ratio. We used 40% because the average rates (= Total income tax/Pretax income) for 1996 and 1995 are not identical, but are both close to 40%.

$$\text{1995 Cash flow coverage ratio} = \frac{51.7 + 0 + 22.7}{0 + 7.7 + \dfrac{1}{(1 - 0.40)} + \dfrac{0}{(1 - 0.40)}} = 7.94$$

$$\text{1996 Cash flow coverage ratio} = \frac{37.9 + 0 + 20.1}{0 + 8.0 + \dfrac{1}{(1 - 0.40)} + \dfrac{0}{(1 - 0.40)}} = 6.00$$

Appendix Exercise continued

Profitability Ratios

Gross profit margin = 260.6/546.9 = 47.7%	238.5/485.9 = 49%
Net profit margin = 25.9/546.9 = 4.7%	18/485.8 = 3.7%
Return on assets = 25.9/580.4 = 4.5%	18/501.1 = 3.6%
Earning power = 51.7/580.4 = 8.9%	37.9/501.1 = 7.6%
Return on equity = 24.9/313.7 = 7.9%	17/293.3 = 5.8%

Market Value Ratios

P/E = 47.25/2.77 = 17x

Earnings yield = 2.77/47.25 = 5.9%

Dividend yield = 0.50/47.25 = 1.1%

Market-to-book ratio = 47.25/34.86 = 1.36x

CHAPTER 3 -- THE FINANCIAL ENVIRONMENT, PRINCIPLES, AND FUNDAMENTAL CONCEPTS

A1. The **opportunity cost** of an alternative is the difference between its value and the value of the best possible alternative.

A2. A **principal-agent relationship** is a relationship in which one person, the agent, is responsible for making decisions that affect another person, the principal.

A3. The problem of **moral hazard** can arise whenever an agent can take unseen actions for personal benefit when those actions are costly to the principal. For example, a real estate agent might inform a potential buyer of the homeowner's lowest acceptable offer in exchange for other benefits.

A4. A **portfolio** is a group or set of securities.

A5. A **zero-sum game** is a situation in which one player can gain *only* at the expense of another player.

A6. A **sunk cost** is a cost that has already been incurred and will not be altered by subsequent decisions.

A7. A firm's announcement of a dividend increase may signal higher expected future earnings for the firm.

A8. The problem of **adverse selection** arises whenever participation in a market is an apparently negative signal that discourages the inclusion of good-quality products in that market. The classic example involves the market for used cars. If buyers expect all used cars to be "lemons," they are only willing to pay a relatively low price for any used car; as a result, the owners of high-quality used cars will not participate in the used car market because they cannot receive a fair price.

A9. **Arbitrage** is the act of buying and simultaneously selling an asset, where the sale price is higher than the purchase price, so that the difference provides a riskless profit.

A10. a A **spot market** is a market to buy and sell something today for immediate delivery. In a **futures market**, assets are bought and sold for future delivery.

b. A **call option** is a right to buy. A **put option** is a right to sell.

c. **Options contracts** confer rights on their buyers and obligations on their sellers. **Futures contracts** confer obligations on both parties.

d. A **broker** is a middleman who helps clients buy and sell securities for a commission, but who does not take ownership of the securities. A **dealer**, on the other hand, maintains inventories of securities, and buys and sells to clients, to and from these inventories.

e. An **investment banker** specializes in marketing new securities in the primary market. Investment bankers facilitate trading in securities. A **financial intermediary** is an institution that invests in securities, but is itself financed through the issuance of other financial claims. Financial intermediaries transform the characteristics of financial claims from one form (for example, savings account) to another (for example, commercial loan).

f. The **primary market** is the market in which newly created securities are sold by a firm to obtain additional financing. The **secondary market** is where previously issued or "used" securities are traded.

g. An **initial public offering (IPO)** involves the sale of a firm's securities to the public for the first time. Any subsequent offerings of the firm's securities are **seasoned offerings.**

h. **Futures contracts** are standardized **forward contracts** that are traded on an exchange. The standardization and exchange trading of the futures contracts makes them more liquid than forward contracts.

i. Shares of **stock** in a firm represent ownership of the firm. **Bonds** issued by a firm are long-term debt obligations of the firm.

A11. An **option** provides its owner a right, without an obligation, to do something. A **call option** is the right to buy, and a **put option** is the right to sell.

A12. a. $FV = PV \times (1+r)^n = 1000 \times (1.10)^1 = \$1100$

b. $FV = PV \times (1+r)^n = 1000 \times (1.10)^5 = \$1610.51$

A13. Now, $r = 0.025$ (the quarterly interest rate) and $n = 4$ for part a and $n = 4 \times (5) = 20$ for part b (n is the number of periods, that is, in this case the number of quarters).

a. $FV = 1000 \times (1+0.025)^4 = \$1103.81$

b. $FV = 1000 \times (1+0.025)^{20} = \$1638.62$

A14. $PV = FV/(1+r)^n = 10{,}000/(1.12)^7 = \$4523.49$

A15 The **term structure of interest rates** is depicted by the yield curve of zero coupon U.S. Treasury securities. It displays the relationship between time to maturity and yield to maturity for those securities. Normally, longer term securities have higher yields. The existence of the term structure is influenced by investor demands for securities of particular maturities, the potential risk of holding longer term securities, and investor expectations about future interest rates. Informally, the phrase *term structure* is used to refer to the relationship between maturity and interest rates, generally.

B1. According to the Principle of Self-Interested Behavior, each party to a financial transaction will choose the course of action that is most financially advantageous to that party, given the information they possess. A particular piece of information may result in a particular action when the decision-maker acts in his own self-interest. In that case, observing the decision-maker's action may allow us to deduce the information that is known to the decision-maker.

B2. The term **efficient capital market** means that the price of any security traded in that market is a function of all available information -- information about the firm, its competitors, the economy, etc. -- and when any new information becomes available, the price of the security quickly changes to its correct price based on the new set of information.

B3. **Limited liability** is a legal concept stating that the financial liability of an asset owner is limited in some manner. For example, a stockholder in a publicly traded firm such as IBM or General Motors cannot be held responsible for a liability of the corporation. If a borrower's financial liability is limited to a particular amount, that limit gives the borrower the option to default and not fully repay the loan.

B4. If Teradyne's business was truly "jumping," we would expect the firm's chairman (as well as other investors) to be buying shares rather than selling them. The fact that the chairman sold $793,600 in shares means that he must have expected those shares to be worth less in the future. The Signaling Principle implies that the chairman's actions conveyed negative information, despite the optimistic words of the firm's executives.

B5. According to the Signaling Principle, an action we observe is the optimal action of a self-interested player, given a particular information set. If a second player expects to face the same (or nearly the same) information set but does not know exactly what is in the information set and it is costly (or impossible) to determine the information with accuracy, the second player can infer that the action taken by the first player will also be in the second player's best interest. As a result, we have the Behavioral Principle -- when all else fails, look at what others are doing for guidance.

B6. Two appropriate applications of the Behavioral Principle are (1) the case where there is a limit to our understanding -- for example, trying to choose an optimal capital structure for your firm when there is no clear-cut financial theory upon which to base your choice; and (2) the case where its use is more cost-effective than obtaining sufficient information to answer the question using the most accurate method -- for example, valuing certain assets according to the prices of similar assets, rather than undertaking the extensive information gathering and estimation needed to make a more accurate valuation. Two inappropriate applications are (1) "blind imitation" that involves following the actions of other firms that face a situation that is quite different from your situation -- for example, matching your competitor's dividend policy when your expected future earnings are significantly different; and (2) "blind imitation" that involves following the actions of the majority when a single, clearly correct, best course of action exists for the problem at hand.

B7. You would want to guard against the **free-rider** problem in any situation where you expend resources to determine a best course of action and others can receive the same benefits by simply imitating your actions. For example, if you have researched and developed a new, cost-saving, way to manufacture automobile tires, you should apply for a patent so that other firms in the tire industry cannot costlessly imitate your new production process.

B8. Interest compounding occurs when an amount is invested for more than one period and you earn "interest on your interest." In the first period, the amount invested earns interest. In subsequent periods, in addition to the investment earning interest, previously earned interest also earns interest.

B9. Let PV_A be the present value of the $2500 installment to be received in 4 years and let PV_B be the present value of the $2500 installment to be received in 5 years. Then the present value of the total $5000, PV, is equal to PV_A plus PV_B.

$$PV = PV_A + PV_B = 2500/(1.10)^4 + 2500/(1.10)^5 = 1707.5336 + 1552.3033 = \$3259.84$$

B10. The reported trading price just after that 1-hour period would be higher than the reported trading price just prior to it.

B11. The Principle of Two-Sided Transactions is important to financial decision-making because when we analyze any transaction, we must remember that someone else is analyzing that same transaction from the opposite point of view. This is much like any game of strategy, such as chess.

B12. A financial decision only has an impact if it changes future outcomes from what they would have been if the decision had not been undertaken. As a result, the appropriate way to value a financial decision is to measure all the incremental changes in future outcomes (incremental benefits) that it will cause. A sunk cost has no effect on the incremental benefit from a decision because it has already been incurred and cannot be changed by any subsequent decisions.

B13. Equity represents ownership interest in a firm, debt does not. Equityholders are entitled to a firm's cash flows only after all more senior obligations have been met. Debt is one type of senior obligation, and it involves a contractual promise on the part of the firm to make interest and principal payments. Because equityholders are not promised regular cash flows, they bear more of the firm's risk. However, equityholders also receive the rewards in the form of increased capital gains and dividends when a firm is successful. Debtholders, on the other hand, receive contractually promised payments, but forgo the opportunity to share in the firm's success.

C1. Assuming that IBM's new "family" was the embodiment of a valuable new idea, creating more capable machines would render existing machines less cost effective. With specialized assets to manufacture (old generation, less capable) computers, the common stock prices of other computer manufacturers would decline. This is because the price that can be charged for the less capable computers will be less and therefore the specialized assets owned by the other manufacturing firms would be worth less, leading to a lower value of their common stock.

C2. The Principle of Self-Interested Behavior implies that investors will make financial transactions in their own best interest. The Principle of Two-Sided Transactions implies that for every share of stock that is sold by one investor, there must be another (self-interested) investor who is willing to buy a share of stock and people buy or sell until the market price reaches what they think is the correct value of each share. Together these two principles suggest that when new information becomes available, investors will act in their own interest, buying or selling stock until the market price adjusts to the new information. This is consistent with the Principle of Capital Market Efficiency. In essence, each party (side) to a transaction applies the Principle of Self-Interested Behavior, thus enforcing the Principle of Capital Market Efficiency.

C3. It is not possible to measure *exactly* the opportunity cost of an alternative unless it is possible to identify and analyze *every* potential alternative decision. Consider a decision regarding the location of a new manufacturing plant. If we analyze, for example, three locations when there are actually a multitude of potential sites, then the "best alternative" may not be one of the three locations being considered, and the opportunity cost of choosing one of these three locations cannot be measured exactly. For many financial decisions it is impossible, or extremely difficult, to explicitly enumerate *all* of the available alternatives, and the cost of doing so would exceed the benefit of measuring the opportunity cost exactly. Nevertheless, the concept of opportunity cost is extremely important even though it may not be possible to measure it exactly in most situations.

C4. The capital markets are the arenas in which a firm's securities are traded and priced. Understanding how capital markets function and the principles under which they operate help a manager to make decisions that maximize the wealth of the firm's shareholders.

C5. A **perfect capital market** is 100% efficient -- no losses due to friction, and no price differences due to inefficiencies. Therefore, a perfect capital market can be defined as one in which there are no arbitrage opportunities.

CHAPTER 4 -- THE TIME VALUE OF MONEY

Note: Solutions to numerical problems are usually presented in the following format: (1) A description of the problem-solving method is provided the first time a particular problem type is encountered or when problem-solving hints may be needed. (2) The solution is provided in equation format. (3) For much of the material, calculator hints are provided in the problem solution or in brackets at the end of the problem.

Note: Four cautions are in order regarding our calculator solutions: (1) Some calculators require that cash flows be entered with opposite algebraic signs. In other words, PV, and FV and/or PMT, must be entered with one as a positive number and the other as a negative number. (2) Our solutions assume the calculator is set to I/Y = 1, one payment per period (see Appendix A at the back of the book). (3) Our solutions assume the calculator is in the end-of-period mode (see Appendix A at the back of the book). (4) Your answers to some of the problems may vary from ours by a few cents due to different rounding assumptions or to minor rounding differences among calculators.

A1. The following time line shows the expected cash flows:

0	1	2	3	4	5	6	7
		500		1200	800	1500	

a. Find the present value of the set of cash flows by applying the Present-Value formula to each cash flow and summing:

$$PV = \frac{500}{(1.07)^2} + \frac{1200}{(1.07)^4} + \frac{800}{(1.07)^5} + \frac{1500}{(1.07)^6}$$

$$PV = 436.71936 + 915.47425 + 570.38894 + 999.51334 = \$2922.10$$

[Calculate each term separately by entering the expected cash flow as the FV, $r = 7$, PMT = 0, and n = appropriate number of years for that cash flow; compute PV.]

b. Given the present value computed in part a, we can compute the future value at any point in time by applying the Future-Value Formula:

$$FV_5 = PV \times (1 + r)^5 = 2922.10 \times (1.07)^5 = \$4098.40$$

PV = 2922.10, $n = 5$ (10 for part c), PMT = 0, $r = 7$; compute FV.

c. $FV_{10} = PV \times (1 + r)^{10} = 2922.10 \times (1.07)^{10} = \$5748.21$

A2. The following time line shows the expected cash flows:

0	1	2	3	4	5	6	7
	1000	1400		900	600		

a. Following the same procedure as in problem A1,

$$PV = \frac{1000}{(1.08)^1} + \frac{1400}{(1.08)^2} + \frac{900}{(1.08)^4} + \frac{600}{(1.08)^5}$$

$$PV = 925.92593 + 1200.27435 + 661.52687 + 408.34992 = \$3196.08$$

b. $FV_5 = PV \times (1+r)^5 = 3196.08 \times (1.08)^5 = \$4696.09$

c. $FV_3 = PV \times (1+r)^3 = 3196.08 \times (1.08)^3 = \$4026.14$

A3. The present value is found by applying Equation (4.2):

$$PVA_n = CF \times \left[\frac{(1+r)^n - 1}{r \times (1+r)^n}\right] = 500 \times \left[\frac{(1.085)^8 - 1}{0.085 \times (1.085)^8}\right] = \$2819.59$$

[PMT = 500, $r = 8.5$, $n = 8$, FV = 0; compute PV.]

A4. The future value is found by applying Equation (4.1):

$$FVA_n = CF \times \left[\frac{(1+r)^n - 1}{r}\right] = 1000 \times \left[\frac{(1.10)^6 - 1}{0.10}\right] = \$7715.61$$

[PMT = 1000, $r = 10$, $n = 6$, PV = 0; compute FV.]

A5. Using Equation (4.1),

$$FVA_5 = 1200 \times \left[\frac{(1.07)^5 - 1}{0.07}\right] = \$6900.89$$

A6. One approach to this problem is to find the present value of the annuity and then find the future value of that lump sum:

$$PVA_7 = 350 \times \left[\frac{(1.10)^7 - 1}{0.10 \times (1.10)^7}\right] = \$1703.94659$$

then, $FV_{10} = 1703.94659 \times (1.10)^{10} = \$4419.60$

[(Step 1) PMT = 350, $n = 7$, $r = 10$, FV = 0; compute PV and store that answer in the calculator memory. (Step 2) $n = 10$, $r = 10$, and PV = value from Step 1 (recalled from memory); compute FV.]

A7. Using Equation (4.2),

$$PVA_6 = 1000 \times \left[\frac{(1.10)^6 - 1}{0.10 \times (1.10)^6}\right] = \$4355.26$$

A8. Using the loan principal as the present value and applying Equation (4.3) [PV = 5000, $n = 36$, $r = 1$, FV = 0; compute PMT]:

$$CF = PVA_{36} \times \left[\frac{r \times (1+r)^n}{(1+r)^n - 1}\right] = 5000 \times \left[\frac{0.01 \times (1.01)^{36}}{(1.01)^{36} - 1}\right] = \$166.07$$

A9. We first find the annual payment using Equation (4.3),

$$CF = 4000 \times \left[\frac{0.09\times(1.09)^4}{(1.09)^4 - 1}\right] = \$1234.67$$

LOAN AMORTIZATION SCHEDULE

Period:	1	2	3	4
Principal (start of period)	$4000.00	$3125.33	$2171.94	$1132.74
Interest for period	360.00	281.28	195.47	101.95
Balance	4360.00	3406.61	2367.41	1234.69
Payment	1234.67	1234.67	1234.67	1234.67
Principal (start of next period)	3125.33	2171.94	1132.74	0.02*

*Not zero due to rounding error.

A10. First, compute the annual payment:

$$CF = 7500 \times \left[\frac{0.20\times(1.20)^4}{(1.20)^4 - 1}\right] = \$2897.17$$

[PV = 7500, $n = 4$, $r = 20$, FV = 0; compute PMT.]

LOAN AMORTIZATION SCHEDULE

Period:	1	2	3	4
Principal (start of period)	$7500.00	$6102.83	$4426.23	$2414.31
Interest for period	1500.00	1220.57	885.25	482.86
Balance	9000.00	7323.40	5311.48	2897.17
Payment	2897.17	2897.17	2897.17	2897.17
Principal (start of next period)	6102.83	4426.23	2414.31	-0-

A11. Applying the present value perpetuity formula, Equation (4.5),

$$PVA_\infty = CF/r = (800)/(0.11) = \$7272.73$$

Note: For several of the problems that follow, recall that, unless otherwise stated, the compounding frequency is assumed to be the same as the payment frequency.

A12. From Equation (4.6), we find the monthly rate which results in a 9% APR, or 0.09/12 = 0.75%. We then solve for the monthly CF, where there are 12×4 = 48 payments:

$$CF = 15{,}000 \times \left[\frac{0.0075\times(1.0075)^{48}}{(1.0075)^{48} - 1}\right] = \$373.28$$

[$n = 48$, $r = 0.75$, PV = 15000, FV = 0; solve for PMT.]

A13. The weekly interest rate, based on a 10% APR is 10/52 = 0.192%. Using this rate we then solve for PVA_{260} using Equation (4.2).

$$PVA_{260} = 100 \times \left[\frac{(1.00192)^{260}-1}{0.00192\times(1.00192)^{260}}\right] = \$20{,}452.78$$

A14. Here, $n = 12\times10 = 120$, and $r = 0.06/12 = 0.5\%$.

$$FV_{120} = 200 \times \left[\frac{(1.005)^{120}-1}{0.005}\right] = \$32{,}775.87$$

A15. Here, $n = 25\times12 = 300$, FV = 0, and $r = 0.075/12 = 0.625\%$. We solve for CF or PMT.

$$CF = 150{,}000 \times \left[\frac{0.00625\times(1.00625)^{300}}{(1.00625)^{300}-1}\right] = \$1108.49$$

A16. For this problem, we are given the APY. Then, using the Present-Value Formula:

$$PV = \frac{10{,}000}{(1.082)^{7.8}} = \$5{,}407.88$$

B1. The following time line shows the expected cash flows:

0	1	2	3	4	5	6	7	8	9	10
			1500	1500	1500	1500	1500	1500	1500	

The problem can be solved as the difference between the present value of a 9-year annuity and the present value of a 2-year annuity:

$$PV = 1500 \times \left[\frac{(1.10)^9 - 1}{0.10\times(1.10)^9}\right] - 1500 \times \left[\frac{(1.10)^2 - 1}{0.10\times(1.10)^2}\right]$$

$$= 8638.5357 - 2603.3058 = \$6035.23$$

B2.

a. $PVA_{10} = 1000 \times \left[\frac{(1.08)^{10} - 1}{0.08 \times (1.08)^{10}}\right] = \$6710.08$

b. $PVA_{20} = 1000 \times \left[\frac{(1.08)^{20} - 1}{0.08 \times (1.08)^{20}}\right] = \$9818.15$

c. $PVA_{50} = 1000 \times \left[\frac{(1.08)^{50} - 1}{0.08 \times (1.08)^{50}}\right] = \$12{,}233.48$

d. $PVA_{100} = \$1000 \times \left[\frac{(1.08)^{100} - 1}{0.08 \times (1.08)^{100}}\right] = \$12{,}494.32$

e. $PVA_{\infty} = (1000)/(0.08) = \$12{,}500.00$

B3. Using the equation for the present value of an annuity and inserting the values from the problem results in:

$$18{,}000 = 439.43 \times \left[\frac{(1 + r)^{48} - 1}{r \times (1 + r)^{48}}\right]$$

This equation cannot be solved algebraically. It must be solved by trial-and-error or by using a computer or calculator. On your calculator enter PV = 18,000, $n = 48$, FV = 0, and PMT = 439.42; then compute r. The result is $r = 0.6667$.

a. $r = 0.6667$ per month

b. $APR = 0.006667 \times 12 = 8\%$

c. $APY = (1.006667)^{12} - 1 = 8.3\%$

B4. Applying Equation (4.7),

$$APY = [1 + APR/m]^{m} - 1 = [1 + 0.15/12]^{12} - 1 = 0.16075 = 16.075\%$$

B5. Applying Equation (4.8),

$$APY = e^{APR} - 1 = e^{0.15} - 1 = 1.16183 - 1 = 16.183\%$$

B6. Using the equation for the present value of an annuity and plugging the values from the problem results in:

$$130{,}000 = 1007.89 \times \left[\frac{(1 + r)^{240} - 1}{r \times (1 + r)^{240}}\right]$$

As was the case in B3 above, this equation cannot be solved algebraically. It must be solved by trial-and-error or by using a computer or calculator. On your calculator enter PV = 130,000, $n = 240$, FV = 0, and PMT = 1007.89; then compute r. The result is $r = 0.5833$.

a. $r = 0.5833\%$ per month

b. APR = 0.005833 × 12 = 7%

c. APY = $(1.005833)^{12} - 1 = 7.23\%$

B7. The present value is found by applying the PV formula in Table 4-5 and replacing $(1 + r)^n$ with $e^{n \times APR}$:

$$PV = FV_n \times [e^{-n \times APR}] = 3400 \times [e^{-3 \times (0.11)}] = 3400 \times (0.718924) = \$2444.34$$

B8. Using the present value annuity formula and replacing $(1 + r)^n$ with $e^{n \times APR}$ yields

$$PVA_5 = CF \times \left[\frac{e^{n \times APR} - 1}{APR \times e^{n \times APR}}\right] = 15{,}000 \times \left[\frac{e^{5 \times (0.12)} - 1}{0.12e^{5 \times (0.12)}}\right]$$

$$= 15{,}000 \times (3.759903) = \$56{,}398.55$$

B9. Here, $n = 31/12 = 2.58333$, therefore, the present value formula gives

$$PV - \frac{FV_n}{(1 + r)^n} = \frac{4500}{(1.10)^{2.58333}} = \$3517.88$$

B10. First, determine the APY.

APY = $e^{0.10} - 1 = 10.517\%$

Second, use the APY as the discount rate to determine the future value of an annuity.

$$FVA_5 = 20{,}000 \times \left[\frac{(1.10517)^5 - 1}{0.10517}\right] = \$123{,}364.92$$

B11. First, calculate the PV of a 5-year annuity with annual payments of \$3000 and an APY of 7%.

$$PVA_5 = 3000 \times \left[\frac{(1.07)^5 - 1}{0.07 \times (1.07)^5}\right] = \$12{,}300.59$$

Treat this amount as if it had been received 9 months ago and compounded for 9/12 of a year.

$12{,}300.59 \times (1.07)^{0.75} = \$12{,}940.88$

B12. Finding the future value of the 4-year annuity will give the value at the end of year 4. To find the value 4.35 years from now, that FV must be compounded for 0.35 years:

$$FVA_4 = 1200 \times \left[\frac{(1.10)^4 - 1}{0.10}\right] = \$5569.20$$

$$FV_{4.35} = 5569.20 \times (1.10)^{0.35} = \$5758.11$$

B13. Computing the present value of an ordinary 6-year annuity will give the value 5 months after the actual first payment at $t = 7$ months. To find the value at t = 0, that PV will have to be compounded for 5 months (5/12 of a year):

$$PVA = 2500 \times \left[\frac{(1.12)^6 - 1}{0.12 \times (1.12)^6}\right] \times (1.12)^{5/12} = \$10{,}775.52$$

B14. r can be computed as:

$$r = \left[\frac{FV_n}{PV}\right]^{1/n} - 1 = \left[\frac{2000}{1423.56}\right]^{1/3} - 1 = 12\%$$

[PV = 1423.56, FV = 2000, $n = 3$, PV = 0; compute r.]

B15. $FV_n = PV \times (1+r)^n$

$FV_n/PV = (1+r)^n$

$\ln(FV_n/PV) = n \times \ln(1+r)$

$n = [\ln(FV_n/PV)]/[\ln(1+r)] = [\ln(1000/592.03)]/[\ln(1.10)] = 0.524198/0.095310 = 5.5$ years

[FV = 1000, PV = 592.03, $r = 10$, PMT = 0; compute n.]

B16. Doubling the PV amount means that $FV_n = 2 \times (PV)$, so we can use the equation derived in problem B15 with $(FV_n/PV) = 2$ or use your calculator with any dollar amounts so that $FV_n = 2 \times (PV)$, for example, $FV_n = \$2$ and PV = \$1:

a. 17.67 years

b. 8.04 years

c. 4.96 years

B17. We need to find the discount rate that solves the following equation:

$$27{,}469.57 = 5000 \times \left[\frac{(1+r)^8 - 1}{r\times(1+r)^8}\right]$$

Because this equation cannot be solved algebraically, it must be solved using trial and error or using your calculator by entering PMT = 5000, n = 8, FV = 0, and PV = 27,469.57; then computing r. The rate is 9.2%.

B18. The problem can be solved as the difference between the present value of an 18-year annuity and a 3-year annuity.

$$PV = 1800 \times \left[\frac{(1.073)^{18}-1}{0.073\times(1.073)^{18}}\right] - 1800 \times \left[\frac{(1.073)^3 - 1}{0.073\times(1.073)^3}\right]$$

$$= 17{,}720.81 - 4698 = \$13{,}022.81$$

Alternatively, the problem can be solved by computing the present value of the 15-year annuity which will give the value 1 year before the first payment at t = 3. To find the value at t = 0, that PV will have to be discounted for 3 years:

$$PV = 1800 \times \left[\frac{(1.073)^{15} - 1}{0.073\times(1.073)^{15}}\right] \times \left[\frac{1}{(1.073)^3}\right] = \$13{,}022.81$$

[(Step 1) PMT = 1800, n = 15, r = 7.3, FV = 0; compute PV and store it in memory. (Step 2) n = 3, r = 7.3, and FV = value recalled from memory; compute PV.]

B19. One approach to this problem is to compute the present value of a 15-year annuity and then subtract the present value of the two missing payments:

$$PV = 1000 \times \left[\frac{(1.12)^{15} - 1}{0.12\times(1.12)^{15}}\right] - \frac{1000}{(1.12)^3} - \frac{1000}{(1.12)^5} = \$5531.66$$

[(Step 1) Compute the annuity PV: PMT = 1000, n = 15, r = 12, FV = 0; compute PV. (Step 2) Compute the two lump sum PVs individually: FV = 1000, n = 3 (then n = 5), r = 12, PMT = 0; compute PV. (Step 3) Subtract the two values computed in Step 2 from the value computed in Step 1.]

B20. a. Monthly interest is computed with your calculator by entering the principal amount, monthly payment, and number of months: PMT = 374.90, n = 360, FV = 0, and PV = 40,000; compute r. The resulting interest rate is r = 0.9% per month.

b. APR = 12×(0.9%) = 10.8% per year.

c. $APY = (1 + APR/m)^m - 1 = (1.009)^{12} - 1 = 0.1135 = 11.35\%$

B21. a. To compute the payments on Chrysler's special-financing deal, set PV = 23,000, n = 42 FV = 0, and r = 0.1833 (= 0.022/12). Then compute PMT = $569.47.

b. To determine whether you are better off with the cash-back offer or with the special-financing, either compare monthly payments under the two plans or the present values under the plans. In either case, choose whichever plan results in the lower value.

To compare the plans on the basis of their respective PV's, set $r = 0.6917$ ($= 8.3/12$), $n = 42$, FV = 0, and PMT = 569.47; then compute PV = \$20,695.43. Since this value is less than the real cash-back price, take the special financing.

To compare the plans on the basis of monthly payments, find the monthly payment for borrowing from the bank at 8.3% and buying the car at the real cash back price. Set $r = 0.6917$ ($= 8.3/12$), $n = 42$, PV = 21,000 (= 23,000 - 2000), FV = 0, and solve for PMT = \$577.86. Since the monthly payment is lower under the special-financing deal, take the special-financing.

B22. First, the payments required by Performance Auto are computed: PV = 31,000, $n = 36$, FV = 0, and $r = 0.0833$ per month (1% APR). This calculation gives the monthly payment of \$874.45. The present value of the payments required by Performance Auto is computed from the discount rate of 1% per month (the opportunity cost) and 36 payments of \$874.45 to yield \$26,327.51. This is more than the \$25,500 cash price so you should borrow the money from Bob's Bank and pay the cash price.

B23. The cash flows are shown in the following time line:

					8000	8000	8000	8000	
-5	-4	-3	-2	-1	0 FR	1 SO	2 JR	3 SR	4
X					X				

The first monthly payment will be made one month after $t = -5$ and the last monthly payment will be made at $t = 0$, as indicated by the X's on the time line. The four withdrawals will begin at $t = 0$. In order to have enough money saved for the four withdrawals, the future value (at $t = 0$) of the five years of monthly payments must be equal to the present value (at $t = 0$) of the four \$8000 withdrawals. First, compute the APY and then find the value at $t = 0$ of the four withdrawals (note that this present value is simply \$8000 plus the present value of a 3-year, \$8000 annuity):

$$\text{APY} = (1 + 0.0045)^{12} - 1 = 5.5357\%$$

$$\text{PV} = 8000 + 8000 \times \left[\frac{(1.055357)^3 - 1}{0.055357 \times (1.055357)^3}\right] = \$29{,}569.31$$

Finally, set the present value just calculated equal to the future value of a 60-month annuity and solve for the required payment.

$$\text{CF} = 29{,}569.31 \times \left[\frac{0.0045}{(1.0045)^{60} - 1}\right] = \$430.38$$

B24. Compute the monthly discount rate and use that rate to calculate the monthly payment.

$$r_m = (1 + \text{APY})^{1/m} - 1 = (1.10)^{1/12} - 1 = 0.79741\%$$

$$CF = 10{,}000 \times \left[\frac{0.0079741 \times (1.0079741)^{36}}{(1.0079741)^{36} - 1}\right] = \$320.65$$

[PV = 10,000, FV = 0, n = 36, r = 0.7974; compute PMT].

B25. In order to have enough money saved to be able to withdraw $20,000 per year during your retirement, the future value (at t = 20) of your savings must equal the present value (also at t = 20) of the $20,000 perpetuity. At an APY of 4%, the present value of the perpetuity is

$$PVA_{\infty} = \frac{20{,}000}{0.04} = \$500{,}000$$

Setting this value equal to the future value and solving for the required payment yields

$$CF = \$500{,}000 \times \left[\frac{0.04}{(1.04)^{20} - 1}\right] = \$16{,}790.88$$

[n = 20, r = 4, PV = 0, and FV = 500,000; compute PMT.]

B26. To compute the payments on Toyota's special-financing deal, set PV = 18,000, n = 36, FV = 0, and r = 0.15833 (= 0.019/12). Then compute PMT = $514.78.

To determine whether you are better off with the cash-back offer or with the special-financing, compare the present values under the plans. Choose whichever plan results in the lower value.

To compare the plans on the basis of their respective PV's, set r = 0.675 (= 8.1/12), n = 36, FV = 0, and PMT = 514.78; then compute PV = $16,403.40. The real cash-back price is 18,000 - 1400 = $16,600, so take the special-financing.

C1. The following time line depicts an n-period **annuity due** with payments of $X per period:

X	X	X	X	. . . . .	X	
0	1	2	3		n-1	n

Ignoring the payment at t = 0, the payment stream comprises an n-1 period annuity. Thus, a simple way to value an annuity due is to compute the value of an n-1 period annuity and add the value of one payment.

a. Using the present value annuity formula for 7 years and adding $750 gives the following result:

$$PVA = 750 + 750 \times \left[\frac{(1.082)^{7} - 1}{0.082 \times (1.082)^{7}}\right] = \$4628.21$$

b. One way to compute the future value of an annuity due is to compute the present value as in part a and then compound that value as a lump sum. This approach is illustrated below:

$$PVA_5 = 400 + 400 \times \left[\frac{(1.104)^5 - 1}{0.104 \times (1.104)^5}\right] = \$1900.9463$$

$$FVA_6 = 1900.9463 \times (1.104)^6 = \$3441.79$$

C2. a. First, we compute the APY with continuous compounding by applying Equation (4.8):

$$APY = e^{\alpha \times APR} - 1 = e^{0.15} - 1 = 16.1834\%$$

Then we compute the monthly rate using Equation (4.7) with our calculator. PV = 1, FV = 1.16834, n = 12, PMT = 0, compute r = 1.25785% per month.

Finally, we compute the payment by applying the cash flow formula, Equation (4.3):

$$CF = 10{,}000 \times \left[\frac{0.0125785 \times (1.0125785)^{108}}{(1.0125785)^{108} - 1}\right] = \$169.81$$

b. The weekly rate is 0.2885% (= 15/52). There are 4.3333 weeks in a month. Thus, the monthly rate is:

$$(1.002885)^{4.3333} - 1 = 1.2562\%$$

Compute the cash flow by again applying Equation (4.3):

$$CF = 10{,}000 \times \left[\frac{0.012562 \times (1.012562)^{108}}{(1.012562)^{108} - 1}\right] = \$169.69$$

C3. a. Computing the present value of the remaining 9 payments will give the value at t = -9 months. To find the value today (at t = 0), that present value will have to be compounded as a lump sum for 9 months.

$$PV = 10{,}000 \times \left[\frac{(1.10)^9 - 1}{0.10 \times (1.10)^9}\right] \times (1.10)^{9/12} = \$61{,}857.65$$

[(Step 1) n = 9, r = 10, PMT = 10,000, FV = 0; compute PV and save that in memory. (Step 2) n = 0.75 (= 9/12), r = 10, PV = value recalled from memory; compute FV.]

b. NPV = PV - Cost = 61,857.65 - 61,825 = $32.65

C4. One way to approach this problem is to compute the APY and compound the present value for 1.75 years:

$$APY = (1 + 0.12/2)^2 - 1 = 12.36\%$$

$$FV_{1.75} = 900 \times (1.1236)^{1.75} = \$1103.60$$

C5. a. When the payments are discrete, we must compute the APY and apply Equation (4.2):

$$APY = e^{0.08} - 1 = 8.3287\%$$

$$PVA_{20} = 5000 \times \left[\frac{(1.083287)^{20} - 1}{0.083287 \times (1.083287)^{20}}\right] = \$47{,}912.83$$

b. When the cash flow is received continuously, we use the present-value annuity formula from Table 4.5:

$$PVA_{20} = CF \times \left[\frac{e^{n \times APR} - 1}{APR \times e^{n \times APR}}\right] = 5000 \times \left[\frac{e^{20 \times (0.08)} - 1}{0.08 \times e^{20 \times (0.08)}}\right] = \$49{,}881.47$$

C6. a. In this case, the payments comprise a discrete annual perpetuity and the present value is computed using Equation (4.5):

$$PVA_{\infty} = \frac{10{,}000}{0.074} = \$135{,}135.14$$

b. The answer is the same with a continuously received cash flow. To see this recall how we derived Equation (4.2). Applying the same procedure to the present-value annuity formula in Table 4.5, we get:

$$PVA_n = CF \times \left[\frac{e^{n \times APR} - 1}{APR \times e^{n \times APR}}\right] = CF \times \left[\frac{e^{n \times APR}}{APR \times e^{n \times APR}}\right] - CF \times \left[\frac{1}{APR \times e^{n \times APR}}\right]$$

$$= \frac{CF}{APR} - \frac{CF}{APR \times e^{n \times APR}}$$

As n grows large, $e^{n \times APR}$ also grows large, so the denominator of the final term grows large. In the limit, as n tends toward infinity, the second term becomes zero.

C7. The cash flows are depicted in the following time line:

-5	-4	-3	-2	-1	0 FR	1 SO	2 JR	3 SR	4
					8000	8000	8000	8000	
X								X	

The first monthly payment will be made one month after $t = -5$ and the last monthly payment will be made at $t = 3$, as indicated by the X's on the time line. The four withdrawals will begin at $t = 0$. In order to have enough money saved for the four withdrawals, the value (at $t = 0$) of the monthly payments must be equal to the present value (at $t = 0$) of the four \$8000 withdrawals. Also note that time $t = 0$ corresponds to the 60th payment of a 96-month annuity.

We know that the value at $t = 0$ of the four withdrawals is \$29,569.31 as computed in problem B23. To find the value of the monthly payments at $t = 0$, we must compute the present value (at $t = -5$ years) of the 96-month annuity and then compound that lump sum for 60 months:

$$FV_{60} = PVA_{96} \times (1 + r)^{60} = CF \times \left[\frac{(1 + r)^{96} - 1}{r \times (1 + r)^{96}}\right] \times (1 + r)^{60}$$

Setting FV_{60} in this equation equal to $29,569.31 (present value of the four withdrawals) and solving for the cash flow yields:

$$CF = FV_{60} \times \left[\frac{r \times (1 + r)^{96}}{(1 + r)^{96} - 1}\right] \times \left[\frac{1}{(1 + r)^{60}}\right]$$

$$= 29{,}569.31 \times \left[\frac{0.0045 \times (1.0045)^{96}}{(1.0045)^{96} - 1}\right] \times \left[\frac{1}{(1.0045)^{60}}\right] = \$290.28$$

C8. Because the payments occur every other year, we need to compute the 2-year rate and use that in Equation (4.5) for the present value of the perpetuity:

$$r_b = (1 + APY)^2 - 1 = (1.12)^2 - 1 = 25.44\%$$

$$PVA_{\infty} = \frac{1000}{0.2544} = \$3930.82$$

C9. Because the payments are made every 4 years, we need to compute the 4-year rate. However, computing the present value at that interest rate will give the value 4 years before the first payment (at $t = 2$ years) so the present value will have to be compounded ahead 2 years to get the value at $t = 0$:

$$r_4 = (1 + APY)^4 - 1 = (1.12)^4 - 1 = 57.3519\%$$

$$PV_{\infty} = \frac{500}{0.573519} \times (1.12)^2 = \$1093.60$$

C10. The future value (at $t = 12$) of the payments you make must equals the present value (also at $t = 12$) of the perpetuity of payments you will receive. The future value of your twelve payments is found using Equation (4.1), and the present value of the perpetuity is found using Equation (4.5). [Note that the value of a perpetuity at $t = n$ is equal to its value at $t = 0$ since there are an infinite number of payments to be received in either case.] Setting these two equations equal to one another and solving for r yields:

$$CF \times \left[\frac{(1 + r)^{12} - 1}{r}\right] = \frac{CF}{r}$$

$$(1 + r)^{12} - 1 = 1$$

$$(1 + r)^{12} = 2$$

$$r = (2)^{1/12} - 1 = 5.946\%$$

C11. Again, the future value of the payments you make must equal the present value of the payments you will receive. The present value (at $t = 300$ months) of the perpetuity of payments you will receive is:

$$PVA_{\infty} = \frac{30{,}000}{0.06} = \$500{,}000$$

To compute the required monthly payments, we first need to compute the monthly rate (r) using Equation (4.7) [PV = 1, FV = 1.06, $n = 12$, PMT = 0; compute r], then apply the cash flow formula, Equation (4.4),

$$r = (1 + \text{APY})^{1/12} - 1 = (1.06)^{1/12} - 1 = 0.486755\%$$

$$CF = 500{,}000 \times \left[\frac{0.00486755}{(1.00486755)^{300} - 1}\right] = \$739.33$$

C12. a. Compute the weekly rate using your calculator with $n = 520$ weeks ($= 10 \times 52$), PMT = 31.73, FV = 0, and PV = 10,000. Compute $r = 0.21158\%$ per week.

b. $\text{APR} = 52 \times (0.21158) = 11.00\%$ APR

c. $\text{APY} = (1.0021158)^{52} - 1 = 11.62\%$

C13. First, we need to compute th APY with continuous compounding by applying Equation (4.8):

$$\text{APY} = e^{\text{APR}} - 1 = e^{0.13} - 1 = 13.88\%$$

Next, we compute the monthly rate using Equation (4.7), with our calculator PV = 1, FV = 1.1388, $n = 12$, PMT = 0. Compute $r = 1.089\%$ per month.

Finally, we compute the payment by applying the cash flow formula, Equation (4.3):

$$CF = 50{,}000 \times \left[\frac{0.01089 \times (1.01089)^{300}}{(1.01089)^{300} - 1}\right] = \$566.48$$

[$n = 300$ months ($= 25 \times 12$), $r = 1.089$, FV = 0, and PV = 50,000; compute PMT.]

CHAPTER 5 -- VALUING FINANCIAL SECURITIES

A1. A **required return** is the return that perfectly reflects the riskiness of the expected future cash flows. It is the minimum return a person must earn to be willing to invest.

A2. An **expected return** is the return that would make the NPV of the investment equal zero. It is the return people expect to earn if they make the investment.

A3. **Coupon payments** are periodic interest payments and the **coupon rate** is the interest rate that is applied to the principal to determine the coupon payment.

A4. The **maturity of a bond** is the date on which the final repayment of principal is made.

A5. The market price of the bond is determined by discounting all of the bond's expected future cash flows at the required return:

$$B_0 = \sum_{t=1}^{19} \frac{31.25}{(1.046)^t} + \frac{1000}{(1.046)^{19}} = \$815.78$$

[Many calculators can compute bond prices in a single operation. Enter the number of time periods, the coupon payment per period, the principal amount, and the required return per period: $n = 19$, PMT = 31.25 (= 0.0625×[1000]/2), FV = 1000, $r = 4.6$ (= 9.2/2). Compute PV.]

A6. Follow the same procedure illustrated in A5:

$$B_0 = \sum_{t=1}^{20} \frac{38.75}{(1.042)^t} + \frac{1000}{(1.042)^{20}} = \$956.60$$

A7. We find the price of this preferred stock by applying the formula for a perpetuity, Equation (4.5), with CF = dividend.

$$\text{Price} = \frac{1.00}{0.03} = \$33.33$$

A8. The **payout ratio** is the firm's cash dividend expressed as a proportion of the firm's earnings.

A9. As in A7 above, we find the price of this stock by applying the formula for a perpetuity, Equation (4.5), with CF = dividend.

$$\text{Price} = \frac{0.60}{0.033} = \$18.18$$

A10. We find the price of the stock by applying the Dividend Growth Model, Equation (5.4):

$$P_0 = \frac{D_1}{r-g} = \frac{5.60}{0.10-0.06} = \$140$$

A11. Use Equation (5.4) with g = 0 to find the value of this no-growth stock.

$$P_0 = \frac{5}{0.14} = \$35.71$$

A12. Applying Equation (5.7):

$$g = (1 - \text{POR}) \times r = (1 - 0.55) \times 0.15 = 6.75\%$$

A13. (1) The firm's earnings vary randomly from year to year and the P/E ratio for any given year may be a poor indication of the expected return on future investments. (2) Earnings may not be accounted for at the time they are realized, so reported earnings for a given year may be higher or lower according to when earnings are recognized. (3) Changes in a firm's accounting procedures can radically change the reported earnings even though the value of the firm is unchanged.

A14. Two important factors that limit the usefulness of the stock valuation model are (1) the availability of information--the model is only useful if we are able to estimate the parameter values; and (2) the information cost--the model is useless if the cost of obtaining sufficiently accurate parameter estimates is too high.

A15. The return actually earned in a given time period. Because of risk, the realized return is disconnected from the expected return.

B1. First, the semiannual rate is found by solving the following equation:

$$1030 = \sum_{t=1}^{31} \frac{42.25}{(1+r)^t} + \frac{1000}{(1+r)^{31}}$$

[n = 31, PMT = 42.25, FV = 1000, PV = 1030; compute r.] [Note once again that some calculators require that the PV be entered with an algebraic sign opposite to that of the FV and PMT. You can think of the price you pay for the bond (the PV) as a cash outflow (negative sign) and the future cash flows (the coupon payments and principal amount) as cash inflows (positive sign).]

The semiannual rate is 4.0533%

$$\text{YTM} = 2 \times (4.0533) = 8.107\%$$

B2. Use the procedure shown in B1, and solve the following equation for the semiannual rate, r = 4.7269%:

$$952.50 = \sum_{t=1}^{22} \frac{43.75}{(1+r)^t} + \frac{1000}{(1+r)^{22}}$$

$$\text{YTM} = 2 \times (4.7269) = 9.454\%$$

B3. Use your calculator to find n that solves the following equation:

$$951.25 = \sum_{t=1}^{n} \frac{46.875}{(1.05)^t} + \frac{1000}{(1.05)^n}$$

[PMT = 46.875, r = 5, FV = 1000, and PV = 951.25; compute n.]

n is approximately 31 periods so the bonds will mature on or about October 9, 2012.

B4. Following the same procedure illustrated in B3, we find that n is approximately 28 periods so the bonds will mature on or about April 9, 2011.

B5. Using semi-annual compounding ($n = 2 \times 4 = 8$), we merely solve for r in Equation (5.1) with CPN = 0.

$$790.09 = \frac{1000}{(1+r/2)^8}, \text{ so } r = 2.99\% \text{ is the semiannual rate}$$

YTM = 2 × 0.0299 = 5.98%

[PV = 790.09, FV = 1000, n = 8, PMT = 0, and solve for r. Multiply the resulting answer by 2.]

B6. Following the same procedure as in B5 above:

$$98.24 = \frac{1000}{(1+r/2)^{50}}, \text{ so } r = 4.75\% \text{ is the semiannual rate}$$

YTM = 2 × 0.0475 = 9.5%

B7. Solving Equation (5.6) for g gives:

$$g = r - \frac{D_1}{P_0} = 0.14 - \frac{4.00}{37.50} = 3.33\%$$

B8. a. Fair prices of the 2-, 5-, 10-, and 20-year bonds are \$968.61, \$930.76, \$886.18, and \$838.99, respectively.

b. Fair prices of the 2-, 5-, 10-, and 20-year bonds are \$951.24, \$893.83, \$828.64, and \$764.06, respectively.

c. Fair prices of the 2-, 5-, 10-, and 20-year bonds are \$986.39, \$969.58, \$949.04, and \$925.78, respectively.

[PMT = 36.25 and FV = 1000 for all calculations; n = 4, 10, 20, or 40 periods; r = 4.5 for part a (YTM = 9%), 5.0 for part b (YTM = 10%), and 4.0 for part c (YTM = 8%). Compute the PV in each case.]

d. The change in bond price (as a percentage of the original price) is larger the longer the maturity of the bond. Thus the longer the maturity of the bond, the more sensitive its price is to changes in the yield to maturity. That is, interest rate risk increases with increased maturity.

B9. Following the same procedure illustrated in B8,

a. Fair prices of the 1-, 7-, and 15-year bonds are \$1012.97, \$1072.62, and \$1118.88.

b. Fair prices of the 1-, 7-, and 15-year bonds are \$1022.56, \$1129.68, and \$1218.41.

c. Fair prices of the 1-, 7-, and 15-year bonds are \$1003.51, \$1019.17, and \$1030.54.

d. Interest rate risk is higher for longer-maturity bonds than for shorter-maturity bonds.

B10. We compute a yield-to-call (YTC) just as we compute a YTM, but we assume the bond is called on the first call date at the call price. For this problem PV = 1107.67, FV = 1080, $n = 8$ (= 2×4), and PMT = 50 (= 100/2). Solving for r results in a 4.245% semiannual rate, which must be multiplied by 2 to give YTC = 8.49%.

B11. Using the reasoning in B10 above, PV = 1112.05, FV = 1050, $n = 4$ (= 2×2), and PMT = 70 (= 140/2). Solving for r results in 5% semiannual rate, which must be multiplied by 2 to give a YTC = 10%.

B12. a. Assuming there are 29 payments remaining, the semiannual rate is computed with your calculator by entering $n = 29$, PMT = 75, FV = 1000, and PV = 1100. The resultant interest rate is 6.7088%. YTM = 13.4176% (= 2×6.7088) and APY = 13.8677% (= $(1.067088)^2 - 1$).

b. Assuming there are 28 payments remaining, the semiannual rate is 6.6998% This corresponds to YTM = 13.3996% and APY = 13.8485%.

B13. a. Assuming there are 13 payments remaining, the semiannual interest rate is computed with your calculator to be 4.846%. Therefore, YTM = 9.692% and APY = 9.927%.

b. Assuming there are 12 payments remaining, the semiannual interest rate is computed to be 4.971%. Therefore, YTM = 9.942% and APY = 10.189%.

B14. Apply the Dividend Growth Model to obtain the following answer.

$$P_0 = \frac{2}{0.14 - 0.07} = \$28.57$$

B15. First, we need to compute the expected earnings per share and dividend in each time period:

$EPS_0 = 0.25$ — $D_0 = 0.10$
$EPS_1 = 0.25\times(1+4.00) = 1.25$ — $D_1 = 0.10$
$EPS_2 = 1.25\times(1+0.75) = 2.1875$ — $D_2 = 0.10$
$EPS_3 = 1.25\times(1+0.75)^2 = 3.828125$ — $D_3 = 0.10$
$EPS_4 = 1.25\times(1+0.75)^3 = 6.69922$ — $D_4 = 0.10$
$EPS_5 = 6.69922\times(1+0.03) = 6.9002$ — $D_5 = 0.10$
$EPS_6 = 6.69922\times(1+0.03)^2 = 7.1072$ — $D_6 = 0.80\times(7.1072) = 5.68576$

Because the firm expects a constant growth in dividends and a constant payout ratio from year 6 onward, we can use the Dividend Growth Model, Equation (5.4), to compute the expected price at year 5:

$$P_5 = \frac{D_6}{r-g} = \frac{5.68576}{0.32-0.03} = \$19.6061$$

Finally, find the price at $t = 0$ by discounting the expected dividends and the expected price at $t = 5$:

$$P_0 = \frac{0.10}{(1.32)} + \frac{0.10}{(1.32)^2} + \frac{0.10}{(1.32)^3} + \frac{0.10}{(1.32)^4} + \frac{0.10+19.6061}{(1.32)^5} = \$5.13$$

B16. We need to compute the expected dividends in each time period:

$D_0 = 1.50$
$D_1 = 1.50$
$D_2 = 1.50$
$D_3 = 1.50$
$D_4 = 1.50$
$D_5 = 1.50\times(1.30) = 1.95$
$D_6 = 1.50\times(1.30)^2 = 2.54$
$D_7 = 1.50\times(1.30)^3 = 3.30$
$D_8 = 3.30\times(1.02) = 3.366$

Because the firm expects a constant growth in dividends from year 8 onward, we can use the Dividend Growth Model, Equation (5.4), to compute the expected price at year 7:

$$P_7 = \frac{D_8}{r-g} = \frac{3.30\times1.02}{0.11-0.02} = \$37.40$$

Finally, the price at $t = 0$ is the present value of the expected future dividends and price at $t = 7$:

$$P_0 = \frac{1.50}{(1.11)} + \frac{1.50}{(1.11)^2} + \frac{1.50}{(1.11)^3} + \frac{1.50}{(1.11)^4} + \frac{1.95}{(1.11)^5} + \frac{2.54}{(1.11)^6} + \frac{3.30+37.40}{(1.11)^7} = \$26.77$$

B17. Applying Equation (5.6) results in:

$$r = \frac{1.80}{25} + 0.02 = 9.2\%$$

B18. First, we need to compute the expected dividend for each time period:

$D_1 = \$1.00$
$D_2 = \$1.00\times(1.25) = \$1.25$
$D_3 = \$1.00\times(1.25)^2 = \$1.5625$
$D_4 = \$1.00\times(1.25)^3 = \$1.953125$
$D_5 = \$1.953125\times(1.04) = \$2.03125$

Given the required return, we would apply the Dividend Growth Model, Equation (5.4), to compute the expected stock price at $t = 4$ and then compute the current price by discounting the expected dividends and the price at $t = 4$. As a result, the required return, r, is the rate that solves the following equation:

$$25 = \frac{1.00}{(1+r)} + \frac{1.25}{(1+r)^2} + \frac{1.5625}{(1+r)^3} + \frac{1.953125}{(1+r)^4} + \frac{\left[\frac{2.03125}{r - 0.04}\right]}{(1+r)^4}$$

The required return is found by trial and error or calculator to be 10.58%.

B19. The expected EPS next year is \$2.56 (= D_1/POR = 1.28/0.5). Using the rearranged Dividend Growth Model gives the dividend growth rate, g:

$$g = r - \frac{D_1}{P_0} = 0.18 - \frac{1.28}{16.00} = 10\%$$

The expected return on future investments, r, is then computed using Equation (5.7).

$$r = g/(1 - \text{POR}) = 0.10/(1 - 0.50) = 20\%$$

C1. We can compute the NPV of the Court Jesters plan by comparing the price per share under current operations with the price per share under the proposed plan.

Current Operations: With expected earnings of \$0.85 and a payout ratio of 75%, the expected dividend next year is \$0.6375 (= 0.75×0.85). Given a growth rate of 5%, we apply the Dividend Growth Model to compute the current share price:

$$P_0 = \frac{D_1}{r-g} = \frac{0.6375}{0.20 - 0.05} = \$4.25$$

Court Jesters: To find the current price under this plan, we first need the schedule of expected dividends. If earnings next year under current operations are expected to be \$0.85 and this corresponds to a 5% growth in earnings, then earnings this year, EPS_0 must be \$0.8095 (= 0.85/1.05). This gives us a starting point to compute expected EPS for each year. Applying the 70% payout ratio in year 4 gives the expected dividend for each year:

$EPS_1 = 0.8095\times(1.17) = 0.947115$	$D_1 = 0$
$EPS_2 = 0.8095\times(1.17)^2 = 1.1081$	$D_2 = 0$
$EPS_3 = 0.8095\times(1.17)^3 = 1.2965$	$D_3 = 0$
$EPS_4 = 1.2965\times(1.065) = 1.3808$	$D_4 = 1.3808\times(0.70) = 0.9665$

Because the constant growth rate of 6.5% begins with D_4, we can apply the Dividend Growth Model to compute the expected price at time $t = 3$:

$$P_3 = \frac{D_4}{r-g} = \frac{0.9665}{0.20 - 0.065} = \$7.16$$

Finally, because there are no dividends before year 4, the current price is simply the present value of the expected price at $t = 3$:

$$P_0 = \frac{7.16}{(1.20)^3} = \$4.14$$

The NPV per share of the Court Jesters plan is the difference between the price per share under that plan and the price per share under current operations:

NPV per share = 4.14 - 4.25 = -\$0.11

C2. Using the rearranged Dividend Growth Model gives the dividend growth rate, g:

$$g = r - \frac{D_1}{P_0} = 0.10 - \frac{2.00}{25.00} = 2\%$$

The expected return, r, is then computed using a rearranged version of Equation (5.7).

$$r = g/(1 - \text{POR}) = 0.02/(1 - 0.80) = 10\%$$

As noted in Section 5.3, in the Dividend Growth Model, price grows at the same rate as dividends, so:

Expected price at time period 4 = $P_4 = P_0 \times (1+g)^4 = 25 \times (1.02)^4 = \$27.06$

C3. From $t = -1$, this bond is equivalent to any 10-period bond that makes interest payments of \$200 at the end of each period and a principal payment of \$1200 at maturity. (Draw a time line to see this.) Because the bond pays interest every other year, the appropriate discount rate is a biannual discount rate. However, computing the present value based on that biannual discount rate will give the value at $t = -1$. To find the value at $t = 0$, we must compound the present value for 1 year (one half of a biannual period). The biannual interest rate, r_b, is the rate that solves the following equation:

$$1100 = \left[\sum_{t=1}^{10} \frac{200}{(1+r_b)^t} + \frac{1200}{(1+r_b)^{10}}\right] \times (1+r_b)^{1/2}$$

Solving by trial and error gives the biannual interest rate of 20.71%. This corresponds to an APY of 9.868% (= $(1.2071)^{½}$ - 1).

C4. Using Equation (5.6) with D_1 = \$2.38, g = 0.06, and P_0 = \$19.45 gives the expected return on the Getty Oil stock of 18.24%. Using the same relationship with D_1 = \$3.35, g = 0.04, and P_0 = \$41.875 gives the expected return on the ConEdison stock of 12%. The different expected returns for the two investments reflects the differing risk of the two investments. The two stocks have different risk-expected return combinations. Phil will only be able to earn a higher return on his investment by taking greater risk.

CHAPTER 6 -- RISK AND RETURN: FINANCIAL SECURITIES

A1. $APY = (1.12 \times 1.07 \times 1.10)^{0.33} - 1 = 9.65\%$

A2. Total Cash Realized = 1070 - 700 = \$370

Realized Return = 370/700 = 52.86%

A3. The two dimensions that make up risk are the variability or uncertainty of future outcomes and the possibility of large negative outcomes.

A4. Risk is defined in the models in this chapter as the standard deviation of the rate of return on total investment. The shortcomings of this definition are that standard deviation measures variability both above and below the mean return and, so, doesn't provide a good measure of the possibility of negative outcomes. It is a useful definition for investment models because return distributions tend to be relatively symmetrical, and because empirical evidence indicates that other measures of risk do not produce better descriptions of returns even though they are more complex and costly to apply.

A5. If two assets are perfectly negatively correlated, then when the return on asset 1 is high, the return on asset 2 will be low and vice versa. If the proportion of each asset is chosen correctly, the high and low returns will always cancel each other exactly so the investment will always earn the same return. If an investment always earns the same return, there is no uncertainty about that return and, therefore, the investment is riskless.

A6. The **portfolio-separation theorem** states that the choice of portfolio to invest in is not based on the investor's attitude toward risk. This is an important finding because it implies that every investor will want to hold the same portfolio of risky assets and will adjust for their own preferences by adjusting the proportion of that portfolio and the proportion of the riskless asset that they hold.

A7. a. The returns on United States Government securities vary over time, so they are not riskless. The value of these securities, like any other bond, moves inversely to the movement of interest rates. There is also some risk, though small, that the government will not make payments on the bonds.

b. The investment is not truly riskless as there is some small positive probability that the United States Government will default in the next three months. As a practical matter, the 90-day Treasury bill is seen as riskless since the probability of default by the government is very low. There are some times, for example the federal budget battle of 1995 and 1996 when the 90-day Treasury bills are not priced as near riskless assets.

c. There is no truly riskless asset.

A8. **The efficient frontier** is the set of attainable portfolios that have the highest expected return for a given level of risk (or equivalently, the lowest risk for a given expected return). By the Principles of Risk Aversion and Self-Interested Behavior, investors will always want to choose investments that maximize their return without increasing risk or that minimize risk without decreasing their return.

A9. The CML (capital market line) portrays the risk-return combinations an investor can achieve by combining an investment in the market portfolio with lending or borrowing at the riskless rate. It is a line on a graph of portfolio expected return versus portfolio standard deviation that touches the y-axis at the riskless rate and is tangent to the efficient frontier. By inspection of the CML, it is clear that one should invest in portfolio M and adjust one's risk/return preferences by borrowing and lending.

A10. The total market value of the securities is the sum of the market values of the four risky assets, or $10,000. The proportion of each investor's portfolio invested in each risky asset is just the value of the assets divided by the total market value.

Proportion invested in asset 1 = 1000/10,000 = 10%

Proportion invested in asset 2 = 2500/10,000 = 25%

Proportion invested in asset 3 = 1500/10,000 = 15%

Proportion invested in asset 4 = 5000/10,000 = 50%

A11. The market portfolio is the best risky portfolio in the economy. It lies on the efficient frontier, and contains every asset available in the market. Using the model in the chapter, all investors will hold a combination of the market portfolio and a riskless asset. Unfortunately, it is difficult to value every asset in the market, which in turn makes it difficult to assign weights and measure returns.

A12. The realized return is the total yield realized on an investment, including dividend yield and capital gains yield.

Realized Return = 6.2% + 15.1% = 21.3%

B1. The fallacy in the statement is that your net worth has, in fact, declined by $10 per share. Although you have not yet recognized for tax purposes the loss, if you wish to sell the stock you will not get that $10 per share.

B2. To create a riskless portfolio set σ_p equal to zero:

$0 = w_1 \times (\sigma_1 + \sigma_2) - \sigma_2$; rearranging, $w_1 \times (\sigma_1 + \sigma_2) = \sigma_2$;

so that $w_1 = \sigma_2/(\sigma_1 + \sigma_2)$; and $w_2 = 1 - w_1 = \sigma_1/(\sigma_1 + \sigma_2)$

B3. Return = (50 + 300)/(3500) = 10.0%

APY = $(1.1)^4 - 1$ = 46.4%

B4. Return = [151.95 + 7×(100) + (42-38)×1.3158 + (42-40)×1.2665]/(3500) = 24.6%

APY = $(1.246)^{1.33} - 1$ = 34.1%

B5. a. Quarter 1 = (3850 - 3500)/3500 = 10%
Quarter 2 = (4103.29 - 3850)/3850 = 6.6%
Quarter 3 = (4359.75 - 4103.29)/4103.29 = 6.3%
Quarter 4 = (4204.04 - 4359.75)/4359.75 = -3.6%

b. (1.1)×(1.066)×(1.063)×(0.964) - 1 = 20.1%

B6. a. $\overline{r}_x = 0.1\times(-10) + 0.3\times(0) + 0.3\times(6) + 0.2\times(10) + 0.1\times(20) = 4.8\%$

$\overline{r}_y = 0.1\times(4) + 0.3\times(8) + 0.3\times(0) + 0.2\times(-5) + 0.1\times(15) = 3.3\%$

b. $\sigma^2_x = 0.1\times(-10 - 4.8)^2 + 0.3\times(0 - 4.8)^2 + 0.3\times(6 - 4.8)^2 + 0.2\times(10 - 4.8)^2 + 0.1\times(20 - 4.8)^2 = 57.76$

$\sigma^2_y = 0.1\times(4 - 3.3)^2 + 0.3\times(8 - 3.3)^2 + 0.3\times(0 - 3.3)^2 + 0.2\times(-5 - 3.3)^2 + 0.1\times(15 - 3.3)^2 = 37.41$

c. $\text{Cov}(X,Y) = 0.1\times(-10 - 4.8)\times(4 - 3.3) + 0.3\times(0 - 4.8)\times(8 - 3.3) + 0.3\times(6 - 4.8)\times(0-3.3) + 0.2\times(10 - 4.8)\times(-5 - 3.3) + 0.1\times(20 - 4.8)\times(15 - 3.3) = 0.16$

d. $\text{Corr}(X,Y) = 0.16/(57.76)^{.5}\times(37.41)^{.5} = 0.00344$

B7. a. $\overline{r}_A = 0.25\times(-15) + 0.15\times(5) + 0.2\times(10) + 0.3\times(25) + 0.1\times(40) = 10.5\%$

$\overline{r}_B = 0.25\times(-10) + 0.15\times(-5) + 0.2\times(0) + 0.3\times(15) + 0.1\times(35) = 4.8\%$

b. $\sigma^2_A = 0.25\times(-15 - 10.5)^2 + 0.15\times(5 - 10.5)^2 + 0.2\times(10 - 10.5)^2 + 0.3\times(25 - 10.5)^2 + 0.1\times(40 - 10.5)^2 = 317.25$

$\sigma^2_B = 0.25\times(-10 - 4.8)^2 + 0.15\times(-5 - 4.8)^2 + 0.2\times(0 - 4.8)^2 + 0.3\times(15 - 4.8)^2 + 0.1\times(35 - 4.8)^2 = 196.19$

$\text{Cov}(A,B) = 0.25\times(-15 - 10.5)\times(-10 - 4.8) + 0.15\times(5 - 10.5)\times(-5 - 4.8) + 0.2\times(10 - 10.5)\times(0 - 4.8) + 0.3\times(25 - 10.5)\times(15 - 4.8) + 0.1\times(40 - 10.5)\times(35 - 4.8) = 236.4$

d. $\text{Corr}(A,B) = 235.6/[(317.25)^{0.5}\times(202.76)^{0.5}] = 0.93$

B8. a. $\overline{r}_M = [3.25 - 3.00 - 4.58 + 1.16 + 1.24 - 2.68 + 3.15 + 3.76 - 2.69 + 2.08 - 3.95 + 1.23]/12 = -0.09\%$

$\overline{r}_{MSFT} = [5.58 - 3.08 + 2.73 + 9.14 + 16.22 - 3.95 - 0.24 + 12.86 - 3.44 + 12.25 - 0.20 - 2.78)/12 = 3.76\%$

b. $\sigma^2_M = [(3.25 + 0.09)^2 + (-3 + 0.09)^2 + (-4.58 + 0.09)^2 + (1.16 + 0.09)^2 + (1.24 + 0.09)^2 + (-2.68 + 0.09)^2 + (3.15 + 0.09)^2 + (3.76 + 0.09)^2 + (-2.69 + 0.09)^2 + (2.08 + 0.09)^2 + (-3.95 + 0.09)^2 + (1.23 + 0.09)^2]/12 = 8.60$

$\sigma^2_{MFST} = [(5.58 - 3.76)^2 + (-3.08 - 3.76)^2 + (2.73 - 3.76)^2 + (9.14 - 3.76)^2 + (16.22 - 3.76)^2 + (-3.95 - 3.76)^2 + (-0.24 - 3.76)^2 + (12.86 - 3.76)^2 + (-3.44 - 3.76)^2 + (12.25 - 3.76)^2 + (-0.2 - 3.76)^2 + (-2.78 - 3.76)^2]/12 = 47.88$

c. Cov(M,MFST) = [(3.25 + 0.09)×(5.58 - 3.76) +(-3 + 0.09)×(-3.08 - 3.76)
+ (-4.58 + 0.09)×(2.73 - 3.76) + (1.16 + 0.09)×(9.14 - 3.76)
+ (1.24 + 0.09)×(16.22 - 3.76) + (-2.68 + 0.09)×(-3.95 - 3.76)
+ (3.15 + 0.09)×(-0.24 - 3.76) + (3.76 + 0.09)×(12.86 - 3.76)
+ (-2.69 + 0.09)×(-3.44 - 3.76) + (2.08 + 0.09)×(12.25 - 3.76)
+ (-3.95 + 0.09)×(-0.20 - 3.76) + (1.23 +0.09)×(-2.78 - 3.76)]/12
= 11.65

d. Corr(M,MFST) = $\{11.65\}/\{(8.60)^{0.5}\times(47.88)^{0.5}\}$ = 0.57

B9. With Corr = -1, Equation (6.9) reduces to

$$\sigma_p = w_1\times(\sigma_1 + \sigma_2) - \sigma_2$$

The combination producing a zero risk portfolio drives the variance of the portfolio to zero.

$$0 = w_1\times(0.07 + 0.09) - 0.09$$

Solving shows w_1 is 0.5625, so 56.25% of the portfolio is invested in asset 1, and the remaining 43.75% of the portfolio is invested in asset 2.

B10. a. Realized Returns Year 1 = [(40 - 25) + 1]/25 = 64%

Realized Returns Year 2 = [(35 - 40) + 1.5)]/40 = -8.75%

Realized Returns Year 3 = [(50 - 35) + 1.75)]/35 = 47.86%

Realized Returns Year 3 = [(75 - 50) + 2.25)]/50 = 54.5%

b. [(75 - 25) + (1 + 1.5 + 1.75 + 2.25)]/25 = 226%

c. Realized APY = $(1 + 2.26)^{0.25} - 1$ = 34.37%

B11. a. $\bar{r}_M$ = [2.43 + 3.61 + 2.73 + 2.80 + 3.63 + 2.13 + 3.18 - 0.03 + 4.01 - 0.50 + 4.10 + 1.74]/12
= 2.49%

$\bar{r}_{HD}$ = [1.63 - 4.01 - 1.39 - 5.65 - 0.30 - 2.10 + 7.98 - 9.09 - 0.31 - 6.58 + 19.13 + 7.61]/12
= 0.58%

b. $\sigma^2_M = [(2.43 - 2.49)^2 + (3.61 - 2.49)^2 + (2.73 - 2.49)^2 + (2.80 - 2.49)^2 + (3.63 - 2.49)^2 + (2.13 - 2.49)^2 + (3.18 - 2.49)^2 + (-0.03 - 2.49)^2 + (4.01 - 2.49)^2 + (-0.50 - 2.49)^2 + (4.10 - 2.49)^2 + (1.74 - 2.49)^2]/12 = 2.01$

$\sigma^2_{HD} = [(1.63 - 0.58)^2 + (-4.01 - 0.58)^2 + (-1.39 - 0.58)^2 + (-5.65 - 0.58)^2 + (-0.30 - 0.58)^2 + (-2.10 - 0.58)^2 + (7.98 - 0.58)^2 + (-9.09 - 0.58)^2 + (-0.31 - 0.58)^2 + (-6.58 - 0.58)^2 + (19.13 - 0.58)^2 + (7.61 - 0.58)^2]/12 = 55.55$

c. $Cov(M,HD) = [(2.43 - 2.49)\times(1.63 - 0.58) + (3.61 - 2.49)\times(-4.01 - 0.58)$
$+ (2.73 - 2.49)\times(-1.39 - 0.58) + (2.80 - 2.49)\times(5.65 - 0.58)$
$+ (3.63 - 2.49)\times(-0.30 - 0.58) + (2.13 - 2.49)\times(-2.10 - 0.58)$
$+ (3.18 - 2.49)\times(7.98 - 0.58) + (-0.03 - 2.49)\times(-9.09 - 0.58)$
$+ (4.01 - 2.49)\times(-0.31 - 0.58) + (-0.50 - 2.49)\times(-6.58 - 0.58)$
$+ (4.10 - 2.49)\times(19.13 - 0.58) + (1.74 - 2.49)\times(7.61 - 0.58)]/12$
$= 5.54$

d. $Corr(M,HD) = \{5.54\}/\{(2.01)^{0.5}\times(55.55)^{0.5}\} = 0.52$

B12. a. It would seem reasonable to assign equal weight to each of the securities analysts' forecasts. There are 24 forecasts so each forecast should have a weight of 1/24. As multiple analysts have chosen each forecast, the probability of the forecasts are

Forecast	Probability
\$2.75	3/24 = 0.125
2.85	2/24 = 0.083
3.00	10/24 = 0.417
3.10	6/24 = 0.250
3.15	3/24 = 0.125

b. $\overline{r}_{SO} = 0.125\times(2.75) + 0.083\times(2.85) + 0.417\times(3) + 0.25\times(3.10) + 0.125\times(3.15) = \$3.00$

c. $\sigma^2_{SO} = 0.125\times(2.75 - 3)^2 + 0.083\times(2.85 - 3)^2 + 0.417\times(3 - 3)^2 + 0.25\times(3.10 - 3)^2 + 0.125\times(3.15 - 3)^2 = \$0.015$

$\sigma_{SO} = (0.015)^{0.5} = \$0.122$

B13. a. $\overline{r}_1 = (0.2)\times(5) + (0.5)\times(9) + (0.3)\times(15) = 10\%$

$\overline{r}_2 = (0.2)\times(-3) + (0.5)\times(4) + (0.3)\times(20) = 7.4\%$

$\overline{r}_3 = (0.2)\times(-7) + (0.5)\times(-1) + (0.3)\times(13) = 2\%$

$\sigma_1 = [0.2\times(5 - 10)^2 + 0.5\times(9 - 10)^2 + 0.3\times(15 - 10)^2]^{0.5} = 3.61$

$\sigma_2 = [0.2\times(-3 - 7.4)^2 + 0.5\times(4 - 7.4)^2 + 0.3\times(20 - 7.4)^2]^{0.5} = 8.66$

$\sigma_3 = [0.2\times(-7 - 2)^2 + 0.5\times(-1 - 2)^2 + 0.3\times(13 - 2)^2]^{0.5} = 7.55$

b. $Cov(R_1,R_2) = 0.2\times(5 - 10)\times(-3 - 7.4) + 0.5\times(9 - 10)\times(4 - 7.4) + 0.3\times(15 - 10)\times(20 - 7.4) = 31$

$Cov(R_1,R_3) = 0.2\times(5 - 10)\times(-7 - 2) + 0.5\times(9 - 10)\times(-1 - 2) + 0.3\times(15 - 10)\times(13 - 2) = 27$

$Cov(R_2,R_3) = 0.2\times(-7 - 2)\times(-3 - 7.4) + 0.5\times(-1 - 2)\times(4 - 7.4) + 0.3\times(13 - 2)\times(20 - 7.4) = 65.4$

$Corr(R_1,R_2) = 31/(3.61)\times(8.66) = 0.99$

$Corr(R_1,R_3) = 27/(3.61)\times(7.55) = 0.99$

$Corr(R_2,R_3) = 65.4/(8.66)\times(7.55) = 1.00$

$\bar{r}_p = 0.25\times(10) + 0.25\times(7.4) + 0.50\times(2) = 5.35\%$

$$\sigma^2_p = (0.25)^2\times(3.61)^2 + (0.25)^2\times(8.66)^2 + (0.5)^2\times(7.55)^2 + 2\times(0.25)^2\times(0.99)\times(3.61)\times(8.66) + 2\times(0.25)\times(0.50)\times(0.99)\times(7.55)\times(3.61) + 2\times(0.25)\times(0.50)\times(7.55)\times(8.66) = 46.71$$

As the correlation coefficients are nearly 1, the perfect positive correlation case provides a close approximation of the actual portfolio variance.

$$\sigma^2_p = [0.25\times(3.61) + 0.25\times(8.66) + 0.5\times(7.55)]^2 = 46.82$$

B14. Just like the Fisher separation theorem (Chapter 4 Appendix), the portfolio separation theorem leads to consensus among the firm's individual investors, even though they may have different attitudes toward risk. The portfolio separation theorem prescribes that all the firm's shareholders will agree that the best risky portfolio to own is the market portfolio and would not want the firm's managers to engage in needless and costly diversification.

B15. Expected Price $= 0.2\times(24) + 0.6\times(33) + 0.2\times(39) = \$32.4$

$\bar{r}_X = (32.4 - 30)/30 = 8\%$

$$\sigma_x = [0.20\times(-0.20-0.08)^2 + 0.60\times(0.10-0.08)^2 + 0.20\times(0.30-0.08)^2]^{0.50} = 16\%$$

B16. a. $\bar{r}_{TB} = 0.1\times(6) + 0.25\times(6) + 0.35\times(6) + 0.15\times(6) + 0.15\times(6) = 6$

$\bar{r}_{GB} = 0.1\times(8) + 0.25\times(7.5) + 0.35\times(7) + 0.15\times(6) + 0.15\times(4) = 6.625$

$\bar{r}_{CB} = 0.1\times(10) + 0.25\times(9) + 0.35\times(8.5) + 0.15\times(6) + 0.15\times(-2) = 6.825$

$\bar{r}_{CS} = 0.1\times(25) + 0.25\times(15.5) + 0.35\times(11.5) + 0.15\times(-1) + 0.15\times(-11.5) = 8.525$

$$\sigma_{TB} = [0.1\times(6-6)^2 + 0.25\times(6-6)^2 + 0.35\times(6-6)^2 + 0.15\times(6-6)^2 + 0.15\times(6-6)^2]^{0.5} = 0$$

$$\sigma_{GB} = [0.1\times(8-6.625)^2 + 0.25\times(7.5-6.625)^2 + 0.35\times(7-6.625)^2 + 0.15\times(6-6.625)^2 + 0.15\times(4-6.625)^2]^{0.5} = 1.234$$

$$\sigma_{CB} = [0.1\times(10-6.825)^2 + 0.25\times(9-6.825)^2 + 0.35\times(8.5-6.825)^2 + 0.15\times(6-6.825)^2 + 0.15\times(-2-6.825)^2]^{0.5} = 3.867$$

$$\sigma_{CS} = [0.1\times(25-8.525)^2 + 0.25\times(15.5-8.525)^2 + 0.35\times(11.5-8.525)^2 + 0.15\times(-1-8.525)^2 + 0.15\times(-11.5-8.525)^2]^{0.5} = 10.778$$

b. $$Cov(TB,GB) = 0.1\times(6-6)\times(8-6.625) + 0.25\times(6-6)\times(7.5-6.625) + 0.35\times(6-6)\times(7-6.625) + 0.15\times(6-6)\times(6-6.625) + 0.15\times(6-6)\times(4-6.625) = 0$$

$$\text{Cov(TB,CB)} = 0.1\times(6 - 6)\times(10 - 6.825) + 0.25\times(6 - 6)\times(9 - 6.825) + 0.35\times(6 - 6)\times(8.5 - 6.825) + 0.15\times(6 - 6)\times(6 - 6.825) + 0.15\times(6 - 6)\times(-2 - 6.825) = 0$$

$$\text{Cov(TB,CS)} = 0.1\times(6 - 6)\times(25 - 8.525) + 0.25\times(6 - 6)\times(15.5 - 8.525) + 0.35\times(6 - 6)\times(11.5 - 8.525) + 0.15\times(6 - 6)\times(-1 - 8.525) + 0.15\times(6 - 6)\times(-11.5 - 8.525) = 0$$

$$\text{Cov(GB,CB)} = 0.1\times(8 - 6.625)\times(10 - 6.825) + 0.25\times(7.5 - 6.625)\times(9 - 6.825) + 0.35\times(7 - 6.625)\times(8.5 - 6.825) + 0.15\times(6 - 6.625)\times(6 - 6.825) + 0.15\times(4 - 6.625)\times(-2 - 6.825) = 4.684$$

$$\text{Cov(GB,CS)} = 0.1\times(8 - 6.625)\times(25 - 8.525) + 0.25\times(7.5 - 6.625)\times(15.5 - 8.525) + 0.35\times(7 - 6.625)\times(11.5 - 8.525) + 0.15\times(6 - 6.625)\times(-1 - 8.525) + 0.15\times(4 - 6.625)\times(-11.5 - 8.525) = 12.959$$

$$\text{Cov(CB,CS)} = 0.1\times(10 - 6.825)\times(25 - 8.525) + 0.25\times(9 - 6.825)\times(15.5 - 8.525) + 0.35\times(8.5 - 6.825)\times(11.5 - 8.525) + 0.15\times(6 - 6.825)\times(-1 - 8.525) + 0.15\times(-2 - 6.825)\times(-11.5 - 8.525) = 38.454$$

$$\text{Corr(TB,GB)} = \text{Corr(TB,CB)} = \text{Corr(TB,CS)} = 0$$

$$\text{Corr(GB,CB)} = 4.684/(1.234)\times(3.867) = 0.982$$

$$\text{Corr(GB,CS)} = 12.959/(1.234)\times(10.778) = 0.974$$

$$\text{Corr(CB,CS)} = 38.454/(3.867)\times(10.778) = 0.923$$

c. $\overline{r}_p = 0.25\times(6 + 6.625 + 6.825 + 8.525) = 6.994$

[Recall that the variance of TB is zero as are the covariances of TB with the other components of the portfolio. Thus, any terms involving TB in the portfolio standard deviation calculation have been dropped.]

$$\sigma_p = [(0.25)^2\times(1.234)^2 + (0.25)^2\times(3.867)^2 + (0.25)^2\times(10.778)^2 + 2\times(0.25)^2\times(4.684) + 2\times(0.25)^2\times(12.959) + 2\times(0.25)^2\times(38.454)]^{0.50} = 3.91$$

d. $\overline{r}_p = 0.20\times6 + 0.20\times6.625 + 0.30\times6.825 + 0.30\times8.525 = 7.13$

$$\sigma_p = [(0.20)^2\times(1.234)^2 + (0.30)^2\times(3.867)^2 + (0.30)^2\times(10.778)^2 + 2\times(0.20\times0.30)\times(4.684) + 2\times(0.20\times0.30)\times(12.959) + 2\times(0.30)^2\times(38.454)]^{0.50} = 4.57$$

B17. a. $\overline{r}_1 = 0.2\times(10) + 0.6\times(15) + 0.2\times(21) = 15.2\%$

$\overline{r}_2 = 0.2\times(8) + 0.6\times(13) + 0.2\times(25) = 14.4\%$

$\overline{r}_3 = 0.2\times(12) + 0.6\times(10) + 0.2\times(8) = 10\%$

$\sigma^2_1 = 0.2\times(10 - 15.2)^2 + 0.6\times(15 - 15.2)^2 + 0.2\times(21 - 15.2)^2 = 12.16$

$\sigma^2_2 = 0.2\times(8 - 14.4)^2 + 0.6\times(13 - 14.4)^2 + 0.2\times(25 - 14.4)^2 = 31.84$

$\sigma^2_3 = 0.2\times(12 - 10)^2 + 0.6\times(10 - 10)^2 + 0.2\times(8 - 10)^2 = 1.6$

$\sigma_1 = (12.16)^{0.5} = 3.49$

$\sigma_2 = (31.84)^{0.5} = 5.64$

$\sigma_3 = (1.6)^{0.5} = 1.26$

b. Project one has the highest expected return, with project two next, and project three following with the least expected returns. Project two has the highest risk followed by project one while project three has the least risk. Project two is dominated by project one. That is you can get both a higher return and lower risk by taking project one rather than project 2. Beyond that, the choice among the different projects depends on your risk/return preference. One possible way to decide is to look at return per unit risk for projects one and three. Take the expected return and divide by the standard deviation for each project.

Project 1 = 15.2/3.49 = 4.36

Project 2 = 14.4/5.64 = 2.55

Project 3 = 10/1.26 = 7.94

The results show you will receive more return per unit risk you are bearing when you choose project 3.

B18. a. $\bar{r}_p = 0.33\times(15.2) + 0.33\times(14.4) + 0.33\times(10) = 13.2\%$

b. $Cov(R_1,R_2) = 0.2\times(10 - 15.2)\times(8 - 14.4) + 0.6\times(15 - 15.2)\times(13 - 14.4) + 0.2\times(21 - 15.2)\times(25 - 14.4) = 19.12$

$Cov(R_2,R_3) = 0.2\times(8 - 14.4)\times(12 - 10) + 0.6\times(13 - 14.4)\times(10 - 10) + 0.2\times(25 - 14.4)\times(8 - 10) = -6.80$

$Cov(R_1,R_3) = 0.2\times(10 - 15.2)\times(12 - 10) + 0.6\times(15 - 15.2)\times(10 - 10) + 0.2\times(21 - 15.2)\times(8 - 10) = -4.38$

$\sigma^2_p = (0.33)^2\times(12.16) + (0.33)^2\times(31.84) + (0.33)^2\times(1.6) + 2\times(0.33)^2\times(19.12) + 2\times(0.33)^2\times(-6.80) + 2\times(0.33)^2\times(-4.38) = 6.70$

$\sigma_p = (6.7)^{0.5} = 2.59$

c. $Corr(R_1,R_2) = 19.12/(3.49)\times(5.64) = 0.97$

$Corr(R_2,R_3) = -6.80/(1.26)\times(5.64) = -0.96$

B19. In this case, security 1 completely dominates security 2. Security 1 has a higher expected return and a lower standard deviation. Therefore, the efficient frontier is made of only security 1:

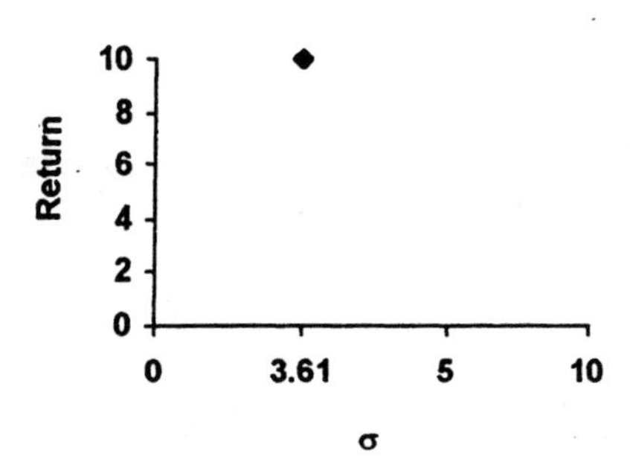

B20. The total market value of the securities is the sum of the market values of the five risky assets, or \$19,100. The proportion of each investor's portfolio invested in each risky asset is just the value of the assets divided by the total market value.

Proportion invested in asset 1 = 1200/19,100 = 6.28%

Proportion invested in asset 2 = 5000/19,100 = 26.18%

Proportion invested in asset 3 = 3400/19,100 = 17.80%

Proportion invested in asset 4 = 2300/19,100 = 12.04%

Proportion invested in asset 5 = 7200/19,100 = 37.70%

So investor Y will invest \$125.60 in asset 1, \$523.60 in asset 2, \$356 in asset 3, \$240.80 in asset 4, and \$754 in asset 5. Likewise, investor X will invest \$62.80 in asset 1, \$261.80 in asset 2, \$178 in asset 3, \$120.40 in asset 4, and \$377 in asset 5.

B21. By the portfolio separation theorem, every investor will choose to own a portion of every asset in the portfolio M and will never choose to hold an asset that is not in the portfolio M. Because there is never an owner for an asset that is not in M, M must contain every asset that has an owner--that is, every asset available in the market.

C1. To show a portfolio of two stocks that are not perfectly correlated can have a lower risk than the less risky asset if Corr < σ_1/σ_2 and $\sigma_1 < \sigma_2$, assume $\rho \approx \rho$ and substitute σ_1 for σ_2 in Equation (6.9). Simplifying after the substitution results in:

$$\sigma^2_p = \sigma^2_1 - 2\times w_1\times\sigma^2_1 + 2\times \text{Corr}(1,2)\times\sigma^2_1$$

Observe that the terms $-2\times w_1\times\sigma^2_1$ and $2\times\text{Corr}(1,2)\times\sigma^2_1$ differ from one another only in regard to w_1 and Corr(1,2). Thus, whenever $w_1 >$ Corr(1,2), $\sigma^2_p < \sigma^2_1$.

C2. Developing new information can be a valuable undertaking even in an efficient market if you are the first person to discover and trade on that information.

C3. It is not possible that any two of these portfolios have a correlation coefficient between them that is equal to minus 1.0. If this were possible, then an appropriate combination of those portfolios would result in a portfolio of zero risk and the set in Figure 6-13 would touch the y-axis in at least one place.

C4. To find the value of w_1 that minimizes the risk of the two-asset investment portfolio, we need to find the value of w_1 that minimizes Equation (6.9). This requires taking the derivative of Equation (6.9) with respect to w_1, setting it equal to zero and solving for w_1. From the chain rule we have:

$$\frac{d\sigma_p}{dw_1} = \frac{1}{2}\left[w_1^2\sigma_1^2 + (1 - w_1)^2\sigma_2^2 + 2w_1(1 - w_1)\text{Corr}\,\sigma_1\sigma_2\right]^{-1/2}[2w_1\sigma_1^2 + 2(1 - w_1)(-1)\sigma_2^2$$

$$+ 2w_1(-\text{Corr}\,\sigma_1\sigma_2) + 2(1-w_1)\text{Corr}\,\sigma_1\sigma_2]$$

Setting the derivative equal to zero gives

$$0 = \frac{\frac{1}{2}\left[2w_1\sigma_1^2-2\sigma_2^2+2w_1\sigma_2^2+2\text{Corr}\,\sigma_1\sigma_2-4w_1\text{Corr}\,\sigma_1\sigma_2\right]}{\left[w_1^2\sigma_1^2+(1-w_1)^2\sigma_2^2+2w_1(1-w_1)\text{Corr}\,\sigma_1\sigma_2\right]^{1/2}}$$

For this equation to equal zero, the numerator must equal zero, so we have:

$$w_1\sigma_1^2 - \sigma_2^2+w_1\sigma_2^2 + \text{Corr}\,\sigma_1\sigma_2 - 2w_1\text{Corr}\,\sigma_1\sigma_2 = 0$$

rearranging: $w_1(\sigma_1^2 + \sigma_2^2 - 2\text{Corr}\,\sigma_1\sigma_2) = \sigma_2^2 - \text{Corr}\,\sigma_1\sigma_2$,

solving for w_1, we have $w_1 = \dfrac{\sigma_2^2 - \text{Corr}\,\sigma_1\sigma_2}{\sigma_1^2 + \sigma_2^2 - 2\text{Corr}\,\sigma_1\sigma_2}$

CHAPTER 7 -- RISK AND RETURN: ASSET PRICING MODELS

A1. Using Equation (7.4),

$\beta_j = Cov(j,M)/\sigma^2_M = 0.0045/0.002 = 2.25$

A2. The market risk premium is the added return required by investors to hold the market portfolio rather than a riskless asset. The premium is just the return on the market portfolio less the return on a riskless asset.

A3. Common stock B has a beta of 1.5. That beta is greater than the beta of common stock A at 1.25. Therefore common stock B has greater beta, or systematic risk, than does common stock A. However, common stock A has a much higher standard deviation than B. Therefore common stock A is said to have greater non-systematic risk than common stock B. In the CAPM it is the systematic risk that determines the stock's risk within a portfolio.

A4.

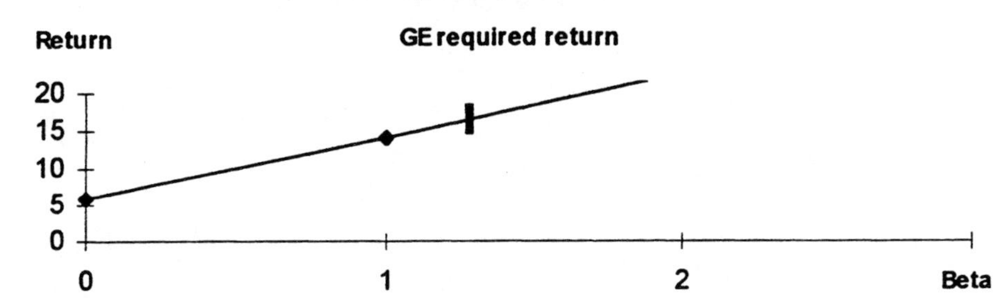

A5. The SML (security market line) is the equation that specifies the required/expected return for a security that is implied by the CML when the market is in equilibrium. It is a line on a graph of security expected return versus security beta that intersects the y-axis at the riskless return and has a slope equal to the excess return on the market portfolio.

The CML describes risk-return possibilities for investment in the market portfolio (along with borrowing or lending at the riskless return). When the capital markets are perfect and in equilibrium, the CML can be inverted to provide a model of the required return for each security in the market; this model is the SML.

A6. **Diversifiable risk** is risk that can be eliminated by further diversification, and **nondiversifiable risk** is risk that cannot be eliminated by further diversification. Nondiversifiable risk is the risk that the asset has if it is held as part of the market portfolio. Diversifiable risk is the portion of the asset's total risk that is not accounted for by nondiversifiable risk.

A7. Investors are compensated for bearing nondiversifiable risk because it cannot be eliminated. Because diversifiable risk can be eliminated easily (and virtually costlessly), investors are not compensated for bearing diversifiable risk.

A8. Using Equation (7.3), $\overline{r}_j = r_f + \beta_j \times (\overline{r}_M - r_f) = 7 + 1.35 \times (12 - 7) = 13.75\%$

A9. An increase in the investor's degree of risk aversion means that the investor requires more return for the same level of risk. Therefore, for any given stock, the investor will require a higher return than the investor required prior to the increase in risk aversion. The investor will also require a greater return for any stock with a positive beta. However, the lower the beta, the lower the required increase in return for the investor.

A10. Domestic projects, within a firm, within an industry, or even across different industries, are subject to many similar risks. The projects may be subject to an economic downturn, changes in the tax code, or a host of other factors. Foreign investment projects have a different set of risks not necessarily strongly correlated with those of the domestic firms. Therefore, foreign investment projects provide diversification benefits to domestic firms. A lower return is required on the international project because of its lower systematic risk.

B1. a.

$$\bar{r}_j = (5.35 - 1.36 - 4.15 + 0 - 1.64 - 8.24 + 4.85 + 1.21 - 4.52 + 9.35 - 2.78 - 0.61)/12 = -0.21$$

$$\bar{r}_M = (3.25 - 3 - 4.58 + 1.16 + 1.24 - 2.68 + 3.15 + 3.76 - 2.69 + 2.08 - 3.95 + 1.23)/12 = -0.09$$

b.

$$\sigma^2_j = [(5.35 + 0.21)^2 + (-1.36 + 0.21)^2 + (-4.15 + 0.21)^2 + (0 + 0.21)^2 + (-1.64 + 0.21)^2 + (-8.24 + 0.21)^2 + (4.85 + 0.21)^2 + (1.21 + 0.21)^2 + (-4.52 + 0.21)^2 + (9.35 + 0.21)^2 + (-2.78 + 0.21)^2 + (-0.61 + 0.21)^2]/12 = 23.52$$

$$\sigma^2_M = [(3.25 + 0.09)^2 + (-3 + 0.09)^2 + (-4.58 + 0.09)^2 + (1.16 + 0.09)^2 + (1.24 + 0.09)^2 + (-2.68 + 0.09)^2 + (3.15 + 0.09)^2 + (3.76 + 0.09)^2 + (-2.69 + 0.09)^2 + (2.08 + 0.09)^2 + (-3.95 + 0.09)^2 + (1.23 + 0.09)^2]/12 = 9.39$$

c.

$$\text{Cov}(j,M) = [(5.35 + 0.21)\times(3.25 + 0.09) + (-1.36 + 0.21)\times(-3 + 0.09) + (-4.15 + 0.21)\times(-4.58 + 0.09) + (0 + 0.21)\times(1.16 + 0.09) + (-1.64 + 0.21)\times(1.24 + 0.09) + (-8.24 + 0.21)\times(-2.68 + 0.09) + (4.85 + 0.21)\times(3.15 + 0.09) + (1.21 + 0.21)\times(3.76 + 0.09) + (-4.52 + 0.21)\times(-2.69 + 0.09) + (9.35 + 0.21)\times(2.08 + 0.09) + (-2.78 + 0.21)\times(-3.95 + 0.09) + (-0.61 + 0.21)\times(1.23 + 0.09)]/12 = 10.16$$

d.

$$\text{Corr}(j,M) = 10.16/(23.52)^{0.5}\times(9.39)^{0.5} = 0.68$$

e. $\beta_j = 1.08$

B2. $\beta_j = 155/125 = 1.24$

$$\bar{r}_j = r_f + \beta_j\times(\bar{r}_M - r_f)$$

$$\bar{r}_j = 5 + 1.24\times(12\text{-}5) = 13.68$$

B3. a. $\bar{r}_{PG} = (5.04 + 2.11 - 0.38 + 5.47 + 2.86 + 0 - 4.17 + 0.73 + 10.99 + 5.19 + 6.79 - 4.05)/12$
$= 2.55$

$\bar{r}_{M} = (2.43 + 3.61 + 2.73 + 2.8 + 3.63 + 2.13 + 3.18 - 0.03 + 4.01 - 0.50 + 4.10 + 1.74)/12$
$= 2.49$

b. $\sigma^2_{PG} = [(5.04 - 2.55)^2 + (2.11 - 2.55)^2 + (-0.38 - 2.55)^2 + (5.47 - 2.55)^2 + (2.86 - 2.55)^2$
$+ (0 - 2.55)^2 + (-4.17 - 2.55)^2 + (0.73 - 2.55)^2 + (10.99 - 2.55)^2 + (5.19 - 2.55)^2$
$+ (6.79 - 2.55)^2 + (-4.05 - 2.55)^2]/12$
$= 18.19$

$\sigma^2_{M} = [(2.43 - 2.49)^2 + (3.61 - 2.49)^2 + (2.73 - 2.49)^2 + (2.8 - 2.49)^2 + (3.63 - 2.49)^2$
$+ (2.13 - 2.49)^2 + (3.18 - 2.49)^2 + (-0.03 - 2.49)^2 + (4.01 - 2.49)^2 + (-0.50 - 2.49)^2$
$+ (4.10 - 2.49)^2 + (1.74 - 2.49)^2]/12$
$= 2.01$

c. $\text{Cov(PG,M)} = [(5.04 - 2.55)\times(2.43 - 2.49) + (2.11 - 2.55)\times(3.61 - 2.49)$
$+ (-0.38 - 2.55)\times(2.73 - 2.49) + (5.47 - 2.55)\times(2.8 - 2.49)$
$+ (2.86 - 2.55)\times(3.63 - 2.49) + (0 - 2.55)\times(2.13 - 2.49)$
$+ (-4.17 - 2.55)\times(3.18 - 2.49)\_+ (0.73 - 2.55)\times(-0.03 - 2.49)$
$+ (10.99 - 2.55)\times(4.01 - 2.49) + (5.19 - 2.55)\times(-0.50 - 2.49)$
$+ (6.79 - 2.55)\times(4.10 - 2.49) + (-4.05 - 2.55)\times(1.74 - 2.49)]/12$
$= 1.57$

d. $\text{Corr(PG,M)} = 1.57/(18.19)^{0.5}\times(2.01)^{0.5}$
$= 0.26$

e. $\beta_{PG} = 0.78$

B4. This is not a valid criticism because asset pricing models are models of *expected* returns and we would predict that (after the fact) actual returns would not necessarily be equal to their expected value.

B5. The beta can be calculated using Equation (7.8):

$$\beta_j = \frac{\text{Cov}(j,M)}{\sigma_M^2} = \frac{\text{Corr}(j,M)\times\sigma_j\times\sigma_M}{\sigma_M^2} = \frac{0.62\times\sqrt{0.1}\times\sqrt{0.0025}}{0.0025} = 3.92$$

B6. Both the CAPM and APT are asset pricing models based on the Principle of Capital Market Efficiency. The CAPM assumes that the mean and standard deviation capture all the necessary information about asset future returns. The APT assumes that the asset's expected return depends on a linear combination of a set of factors and models the expected returns to reflect an absence of arbitrage opportunities. Despite the differences in the development of these two models, when the APT is applied in practice it appears to be a multiple factor extension of the CAPM, and the two alternative models describe actual stock returns with about the same accuracy.

B7. Both statements are correct. σ measures the total risk of a particular asset, while β measures the risk of that asset when it is held in a diversified portfolio -- the nondiversifiable risk.

B8. A zero-beta asset has zero covariance with the market so it has zero *non*diversifiable risk. However, a zero-beta asset does not necessarily have a zero σ. Therefore, it may have diversifiable risk and its total risk is not necessarily zero. A zero-beta asset is riskless in the sense that it will not add any risk to a well-diversified portfolio. Again, however, it certainly may be risky to hold in isolation, with no other investment.

B9. Solving Equation (7.3) for the expected return on the market portfolio, $\bar{r}_M$, gives:

$$\bar{r}_M = \frac{\bar{r}_j - r_f}{\beta_j} + r_f = \frac{17-7}{1.4} + 7 = 14.14\%$$

B10. Because a share of stock is more valuable to a diversified investor than it is to a nondiversified investor, diversified investors are willing to pay a higher price for the share and outbid the nondiversified investors in a competition for the share. As a result, the price that is observed in the market for the share is always the price paid by the diversified investor.

B11. First compute the beta for the asset using Equation (7.4), then use Equation (7.3) to compute the stock's required return:

$$\beta_j = \frac{\sigma_{j,M}}{\sigma_M^2} = \frac{\text{Corr}(j,M) \times \sigma_j \times \sigma_M}{\sigma_M^2} = \frac{0.46 \times \sqrt{0.12} \times \sqrt{0.0016}}{0.0016} = 3.984$$

Then $r_j = 0.06 + 3.984 \times (0.11 - 0.06) = 0.2592 = 25.92\%$

B12. Solving for beta gives:

$$\beta_j = \frac{Corr(j,M) \times \sigma_j \times \sigma_M}{\sigma_M^2} = \frac{Corr(j,M) \times \sigma_j}{\sigma_M} = \frac{0.2 \times (25)}{4} = 1.25$$

Next, substitute the values into Equation (7.3) and solve for the riskless return:

$15 = r_f + 1.25 \times (14 - r_f) = r_f + 17.5 - 1.25 \times r_f$; so that $0.25 \times r_f = 2.5$, and $r_f = 10\%$

B13. Use the information about Stock A to compute the riskless return as above in B12 and then use that value to compute the required return for Stock B:

$15 = r_f + 2.0 \times (10 - r_f) = 20 - r_f$, so that $r_f = 5\%$

and $\bar{r}_B = r_f + \beta_B \times (\bar{r}_M - r_f) = 5 + 1.4 \times (10 - 5) = 12\%$

B14. a. A highly risk-averse investor has utility curves with greater curvature than those found in Figure 7-1. When a utility curve has greater curvature, the point of tangency to any straight, positively sloped line is closer to the y-axis than before. Given the positive, and constant slope of the CML, the new utility curve is tangent to the CML to the left of M.

b. A slightly risk-averse investor has utility curves with less curvature than those found in Figure 7-1. When a utility curve has less curvature, the point of tangency to any straight, positively sloped line is further from the y-axis than before. Given the constant slope of the CML, the new utility curve is tangent to the CML to the right of M.

B15. If the collective degree of risk aversion increases, investors will want more return per unit risk than before. They will, therefore, require a higher return for holding the market portfolio and the market risk premium will increase. The SML will rotate upward around the fixed point found at the riskless return. The SML will rotate around the riskless return because investors bear no more risk and therefore require no more return than before for the riskless asset. The slope of the SML will increase as it equals the market risk premium that has increased.

B16. $\beta_p = w_1\times\beta_1 + w_2\times\beta_2 + w_3\times\beta_3 + w_4\times\beta_4 + w_5\times\beta_5$
$= 1.15\times(0.20) + 0.85\times(0.10) + 1\times(0.15) + 1.30\times(0.25) + 0.80\times(0.30)$
$= 1.03$

B17. a. $\overline{r}_{CG} = 0.20\times(-15) + 0.35\times(15) + 0.30\times(25) + 0.15\times(35) = 15\%$

$\overline{r}_M = 0.20\times(-10) + 0.35\times(10) + 0.30\times(15) + 0.15\times(25) = 9.75\%$

$\sigma^2_M = 0.20\times(-10 - 9.75)^2 + 0.35\times(10 - 9.75)^2 + 0.30\times(15 - 9.75)^2 + 0.15\times(25 - 9.75)^2$
$= 121.19$

$\text{Cov(CG,M)} = 0.20\times(-10 - 9.75)\times(-15 - 15) + 0.35\times(10 - 9.75)\times(15 - 15)$
$+ 0.30\times(15 - 9.75)\times(25 - 15) + 0.15\times(25 - 9.75)\times(35 - 15)$
$= 180$

b. $\beta_{CG} = 180/(121.19) = 1.49$

$\overline{r}_{CG} = 6 + 1.49\times(9.75 - 6) = 11.59\%$

B18. a. $\overline{r}_P = (-5.3 + 13.2 + 6.1 + 2.1 - 8.8 + 15.7 + 3.9 + 12.6 - 7.3 + 10.2)/10 = 4.24$

$\overline{r}_M = (-10.2 + 5.8 + 12.2 - 7.3 - 1.5 + 10.5 + 8.3 + 15.7 - 2.1 + 8.6)/10 = 4$

$\sigma^2_M = [(-10.2 - 4)^2 + (5.8 - 4)^2 + (12.2 - 4)^2 + (-7.3 - 4)^2 + (-1.5 - 4)^2 + (10.5 - 4)^2$
$+ (8.3 - 4)^2 + (15.7 - 4)^2 + (-2.1 - 4)^2 + (8.6 - 4)^2\,]/10$
$= 68.61$

$\text{Cov(P,M)} = [(-10.2 - 4)\times(-5.3 - 4.24) + (5.8 - 4)\times(13.2 - 4.24) + (12.2 - 4)\times(6.1 - 4.24)$
$+ (-7.3 - 4)\times(2.1 - 4.24) + (-1.5 - 4)\times(-8.8 - 4.24) + (10.5 - 4)\times(15.7 - 4.24)$
$+ (8.3 - 4)\times(3.9 - 4.24) + (15.7 - 4)\times(12.6 - 4.24) + (-2.1 - 4)\times(-7.3 - 4.24)$
$+ (8.6 - 4)\times(10.2 - 4.24)]/10$
$= 53.14$

$\beta_P = 53.14/(68.61) = 0.77$

b.

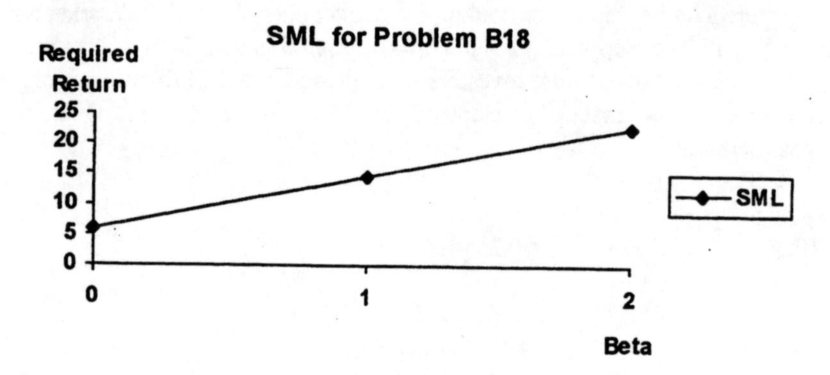

$$\bar{r}_P = 6 + 0.77 \times (8.4) = 12.47\%$$

c. Given the firm's systematic risk, the required return is 12.47%. If an investor expects to get more than that return, they should invest. In this case they expect to earn 19%, so they should invest.

B19. The market value of Not-so-Swift Meat Processors is 75,000,000×$25 = $1,875,000,000. The size premium suggested by Ibbotson Associates for a firm of this size is 1.31%.

$$\bar{r}_N = r_f + \beta_N \times (\bar{r}_M - r_f) + s = 6.5 + 0.95 \times (8.4) + 1.31 = 15.79\%$$

B20. a.

$$V_d = \frac{D_1}{\bar{r}_j - g} = \frac{1.5}{0.15 - 0.10} = 30$$

b.

$$V_n = \frac{D_1}{\bar{r}_{jn} - g} = \frac{1.5}{0.18 - 0.10} = 18.75$$

B21. a. $\bar{r}_R = 6 + 1.15 \times (14 - 6) = 15.2\%$

b. $\bar{r}_R = 6 + 1.15 \times (14 - 6) + 0.50 \times (4 - 6) + 0.75 \times (3 - 6) + 0.25 \times (5 - 6) = 11.7\%$

C1. [See also B8 above.] Beta measures the risk of the asset when it is held in a diversified portfolio, so the measure of risk for the individual asset depends on how the risk of the portfolio changes when that asset is added to the portfolio. When added to the portfolio, a negative beta asset decreases the risk of the portfolio. In this sense, the asset does have negative risk. An investor would hold an asset like this one because of its ability to decrease the risk of the portfolio. Note, however, that if this asset were held alone, its diversifiable risk would cause it to have positive (and perhaps substantial) risk.

C2. Developing a model that accurately values securities provides more benefits than just a means of estimating the value of assets currently traded in the market. If we can identify the determinants of a required return, then we have a basis for predicting future expected returns and for valuing assets that are not currently traded in the capital markets.

C3. It is important to recall that a zero NPV does not mean a zero return on the investment. It just means that in an efficient market, the return that is earned will be commensurate with the risk of the investment --in accordance with the Principle of Risk-Return Trade-Off. The reason for investing in financial securities is to earn that (nonzero) return.

C4. a.

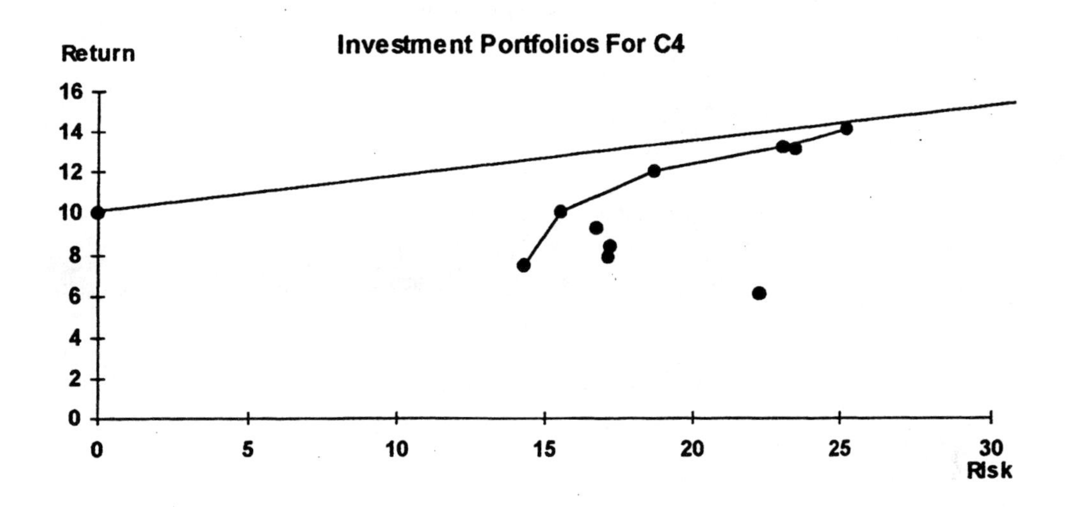

b. The tangent portfolio appears to be portfolio 7 with a 14.1% return and 25.2% risk. The slope of the line through the tangent portfolio is 14.1% return for 25.2% risk, or 0.1627.

c. If you desire a standard deviation of 10%, you should invest in a combination of the riskless asset and portfolio 7. The expected return should be 11.627% (= 10% + 10×[0.1627]). The amount of the portfolio in the risky asset is 10/25.2 = 39.68%. The other 60.32% is in the riskless asset.

d. If you desire a standard deviation of 30%, you should invest in a combination of the riskless asset and portfolio 7. The expected return should be 10% + 30×(0.1627) = 14.881%. The amount of the portfolio in the risky asset is 30/25.2 = 119.05%. The riskless asset should be shorted at 19.05% of the portfolio.

Appendix Exercises

A1. Various researchers have found factors, other than beta, that explain some average stock returns. These discoveries seem to contradict the CAPM claim that beta should subsume all relevant risks. The relevant factors include firm size, leverage, book-to-market, and earnings-price ratio. Of course, empirical tests have their own limitations like a limited number of assets, or use of realized returns. The questions, therefore, may arise from insufficient empirical methodology rather than any flaws in the model.

A2. The empirical evidence suggests average returns are lower for a larger firm and relatively higher for small firms. Indeed, some studies suggest it is size, rather than beta, that is most effective at explaining returns. It appears that size is a proxy for risk.

Appendix Exercises continued

A3. The basic CAPM model can be adjusted for size by adding a size premium variable to the right-hand side of the equation as in Equation (7A.1). The premium depends on the firm size. Stocks with an equity value of below $149 million have a size premium of 4.02%. The premium drops to 2.12% for firms above $149 million and below $617 million in capitalization, and by 1.31% for stocks between $617 million and $2570 million in capitalization.

A4. The observed correlation between measures of leverage and common stock return suggests that measured betas do not include all the effects of high leverage.

B1. Assuming the price on Northern Bearings, Inc. remains at $31.50 and the firm has 50 million shares, the total market capitalization is $1575 million. Therefore, the size premium is 1.31%. This 0.79% less than the low-cap stock premium used on page 7-50. Therefore, the project's required return is 20.6% - 0.79% = 19.81%.

B2. $\overline{r}_j = r_f + \beta_j \times (\overline{r}_M - r_f) + s_j = 6 + 1.2 \times (13 - 6) + s_j = 15.7\%$

$s_j = 1.3\%$

C1. Rather than having a few finite categories, the number of categories could be increased to allow finer gradations. Indeed, the model could be used to develop a size estimate variable rather than just categories. The size adjustment estimates could be updated using current data, or the data could be based on more frequent portfolio rebalancing, weekly or daily rather than quarterly.

CHAPTER 8 -- OPTIONS: VALUING CONTINGENCIES

A1. An **option** is the right, without obligation, to buy or sell an asset. A **call option** is an option to buy, and a **put option** is an option to sell.

A2. The **strike price** is the price at which the holder may buy or sell the asset when the option is exercised. The **exercise value** is the amount of advantage an in-the-money option provides over buying or selling the underlying asset in the market. A call option is **in-the-money** when the price of the underlying asset exceeds the strike price. That is, a call is in-the-money when there is an advantage to exercising over buying the asset in the open market. A call option is **out-of-the-money** when the price of the underlying asset is below the strike price (no advantage to exercise). A put option is in-the-money when the price of the underlying asset is below the strike price. A put option is out-of-the-money when the price of the underlying asset is above the strike price.

A3. Three situations involving "hidden" options are mortgage refunding options, tax-timing options, and capital investment options such as the ability to postpone, expand, or abandon an investment project.

A4. Limited liability makes shares of common stock like a call option on the firm's assets because the shareholders only have to pay the debt (exercise the call) when firm value exceeds the required debt payment. If the value of the firm is less than the required debt payment, the shareholders default on the debt (fail to exercise the call) and the debtholders will own the assets of the firm.

A5. Auto insurance can be viewed as a put option because when an auto is stolen or seriously damaged, the owner can exercise his option to "sell" the auto to the insurance company for the insured amount.

A6. An American call option traded in an efficient capital market would never be worth less than its exercise value because an American option can be exercised at any time during its life and, if it were ever worth less than its exercise value, a person could purchase the option and immediately exercise it to earn a riskless arbitrage profit.

B1. Tom's net investment in the building is $47,800 (= 50,000 - 2200). The outcomes for Tom and Sarah (in $ thousands) are:

Building Value	Tom's Gain (loss)	Tom's Return	Sarah's Gain (loss)	Sarah's Return
$65	$6.20	12.97%	$8.80	400.00%
60	6.20	12.97	3.80	172.73
55	6.20	12.97	-1.20	-54.55
50	2.20	4.60	-2.20	-100.00
45	-2.80	-5.86	-2.20	-100.00
40	-7.80	-16.30	-2.20	-100.00
35	-12.80	-26.78	-2.20	-100.00

B2. a. The exercise value of an option depends on its strike price and the current market value of the underlying asset. For a given strike price, a call's value increases as the value of the underlying asset increases, while a put's value decreases as the value of the underlying asset increases. In other words, the more in-the-money an option is, the greater is the option's value.

b. The parameters which determine an option's time premium are the time until expiration, the risk of the underlying asset, and the market riskless return. The greater the time until expiration, the greater an option's value. The greater the risk of the underlying asset, the greater an option's value. The greater the riskless return, the greater a call's value; however, the greater the riskless return, the lower a put's value.

B3. Jenny's net investment is in the building is \$1.85 million (= 1.5 million + .35 million). Jenny's and Jimmy's outcomes are (in \$ millions).

Building Value	Jenny's Gain (loss)	Jenny's Return	Jimmy's Gain (loss)	Jimmy's Return
\$2.25	\$0.40	21.62%	\$0.35	100.00%
2.00	0.15	8.11	0.35	100.00
1.75	-0.10	-5.41	0.35	100.00
1.50	-0.25	-13.51	0.25	71.43
1.25	-0.25	-13.51	0.00	0.00
1.00	-0.25	-13.51	-0.25	-71.43
0.75	-0.25	-13.51	-0.50	-142.86

B4. An option's time premium is greatest when the strike price equals the underlying asset value because, at that point, the uncertainty is the highest as to whether the option will expire in- or out-of-the-money. For an out-of-the-money option, the option's value is due to the possibility that it will be in-the-money before it expires, and the closer the value of the underlying asset is to the strike price, the greater the probability that the option will become in-the-money before its expiration. For an in-the-money option, the time premium is due to a combination of (1) the possibility that it will be even deeper in-the-money before it expires, and (2) the forgone opportunity of earning the time premium of money on the exercise value (that is, exercising the option now and reinvesting the proceeds). As the value of the underlying asset moves farther from the strike price, the forgone time premium of money on the exercise value grows, but the value of the incremental gain from becoming deeper in-the-money does not grow. Thus, the time premium of an in-the-money option decreases (increases) as the value of the underlying asset moves away from (toward) the strike price.

B5. The time premium for an out-of-the-money American option can be never negative because option values can never be negative. An in-the-money American option's time premium will never be negative because a person could buy the option and immediately exercise it for an arbitrage profit. Market forces eliminate such arbitrage opportunities.

B6. Common stock can be viewed as a put option on the firm's assets with a strike price of \$0. If the stockholders do not want to make a debt payment, they can sell/put the firm's assets to the debtholders at the strike price of \$0.

B7. It is generally better to sell rather than exercise an American option because its total value is its exercise value *plus* its time premium. The general condition that causes an exception to this rule is when the time premium of the option is less than the present value of cash flows that could be received by exercising.

B8. Common stock can be viewed as a series of European call on the firm's assets because shareholders have the right to buy/call the firm from the debtholders whenever an interest or principal payment is due. That is, shareholders can pay the interest or principal (the strike price) and retain ownership of the firm, or fail to pay and relinquish ownership to the debtholders.

B9. An American option will always be at least as valuable as a comparable European option because an American option has the additional right of exercise before the maturity date and also because an option can never have a negative value.

B10. The statement is partly true and partly false. It is true that if the asset goes up in value you get some of the increase but with a smaller initial investment than if you had purchased the underlying asset. It is not true that you do not lose any money if the asset goes down in value, because you do lose the price you paid for the option and the opportunity to have invested that money elsewhere. Owning the option is never less risky than owning the underlying asset. And, as with other financial investments, the price of the option reflects the risk and expected return of the investment.

B11. First compute the probabilities of the 2 outcomes if the return on the business must equal the riskless return of 6%.

6% = (probability of an increase)×(25%) + (1 - probability of an increase)×(-15%)

Solving for the probability of an increase, we get 52.5%. The probability of a decrease is 47.5% (= 1 - 0.525).

Given the 2 outcomes, the strike price of $11 million, and current business value of $10 million, the option can have only two exercise values. These are $0 and $1.5 million (= 1.25×10 million - 11 million). Therefore, the expected value of the outcome is:

0×(0.475) + 1.5 million×(0.525) = $787,500

Finally, the value of call option on the business is the present value of the expected value of the outcome, or

(787,500)/(1.06) = $742,924.53

B12. A firm that has no debt and has only common stockholders can be viewed as a call option in the following way. Because the common stockholders have claim to the entire return distribution, it is as though the strike price is zero and therefore the stockholders *always* exercise their option (by paying the zero strike price) to "reclaim" the assets. Although the no-debt scenario may seem to be stretching the stock-as-an-option characterization, the valuation is valid.

B13. An option is a claim to the "good" outcomes without an obligation to take the "bad" outcomes. Because of diversification, some of the "good" outcomes that could be claimed by the owner of the portfolio of options could not be realized by the owner of the option on the portfolio of assets, since some of the "bad" outcomes cancel them out.

B14. First compute the probabilities of the 2 outcomes if the return on the land must equal the riskless return of 7%.

7% = (probability of an increase)×(15%) + (1 - probability of an increase)×(-8%)

Solving for the probability of an increase, we get 65.22%. The probability of a decrease is 34.78% (= 1 - 0.6522).

Given the 2 outcomes, the strike price of $215,000 and current land value of $200,000, the option can have only two exercise values. These are $0 and $15,000 (= 1.15×200,000 - 215,000). Therefore, the expected value of the outcome is:

0×(0.3478) + 15,000×(0.6522) = $9783

Finally, the value of the call option on the land is the present value of the expected value of the outcome, or

(9783)/(1.07) = $9142.99

C1. This complex bond can be duplicated by purchasing an otherwise identical bond with 20 years to maturity and no early redemption option and purchasing a European put option on the bond with 10 years to maturity and a strike price equal to the face value of the bond. Then, after 10 years, if the value of the bond is less than the face value, you would exercise your put option to sell the bond for its face value. If the value of the bond at $t = 10$ is greater than the face value, you would allow the put option to expire and still have the bond for the remaining 10 years. Alternatively, this complex bond also can be duplicated by purchasing an otherwise identical bond with 10 years to maturity and purchasing a European call option on another identical bond with 10 years to maturity and a strike price equal to the face value of the bond. Then, after 10 years, if the value of the bond is less than the face value, you would fail to exercise your call option and cash in your bond for its face value. If the value of the bond at $t = 10$ is greater than the face value, you would exercise the call option by accepting the replacement 10-year bond in lieu of the face value, thereby still having the bond for the remaining 10 years.

C2. When there is no chance that the option will be out-of-the-money in the remaining time until maturity, the value of the European put is the present value of the exercise value, $(S - P_0)e^{-T \times APR}$. In contrast, because the American option can be exercised now, its value is $S - P_0 > (S - P_0)e^{-T \times APR}$ for any $T > 0$.

C3. Assume that the values of the assets shown on the balance sheet (a portfolio of real estate loans) are in fact fair market values and that the market value of the liabilities are not in excess of that shown on the balance sheet. Purchase of the thrift institution would be essentially the purchase of a series of call options on the asset portfolio, with strike prices and maturities equal to the debt payment obligations. The value of this investment is thus the value of a call option on a portfolio of loans. If interest rate movement causes the value to increase, the investor gets the net increase in value. If interest rate movement causes the value to decrease (all of the loans are defaulted on in the extreme and the "new" institution collapses), the investor has the option to "walk away." If you make up a few simple hypothetical examples to show the effect of interest rate movements on a $1 billion portfolio of loans, they will quickly convince you of the potential value in this investment. For example, assume that the loan portfolio provides income of $140 million per year in perpetuity and that the required return on the portfolio is 14%. Then the portfolio is currently worth $1 billion (= 140 million/0.14). A decline in the required return to 13.8% would net the optionholder an increase of about $14.5 million in the present value of the portfolio (= 140 million/0.138 = 1014.493 million).

C4. A "synthetic" loan could be created which consists of the following portfolio of options that has been sold to a lender for $250,000: 6 put options on the asset, with times to maturity from 1 to 6 years and each with a strike price of $295,001; 5 call options on the asset, with times to maturity from 1 to 5 years and each with a strike price of $250,001; and 1 call option with a 6-year maturity and a strike price of $1. At the end of each of years 1 through 5, the lender would purchase the asset by exercising his maturing call option for $250,001, then immediately sell the asset back by exercising his put option for $295,001, thereby netting interest of $45,000 each year. At the end of year 6, the difference between the strike prices nets the lender $295,000, representing the interest plus the principal.

C5. a. To find the 30-year T price on a calculator, set PMT = 35, FV = 1000, $r = 3$, $n = 60$, and solve for PV = $1138.38. For the contingent amount P, the 30-year T price must be expressed as a percentage of face value, or 113.838.

b.

$$P = \left[1 - \frac{\frac{103 \times 0.06}{0.0488} - 113.838}{100} \right] \times [\$30 \text{ million}] = \$26,159,597$$

c. As stated in the text, the bank's principal obligation is the smaller of $30.6 million or P. So the bank's obligation is $26,159,597.

$$P = \left[1 - \frac{\frac{103 \times 0.08}{0.0488} - 88.688}{100} \right] \times [\$30 \text{ million}] = \$5,950,662$$

d. To calculate P when the yield is 8%, we must first recalculate the 30-year T price with $r = 4$. This results in a price of $886.88, or, expressed as a percentage of par = 88.688.The bank's principal obligation when yields are 8% is $5,950,662.

e. The first term in the numerator of P increases when yields increase while the second term (the 30-year T price) decreases. The result of this is to increase the fraction being subtracted from 1 inside the first brackets in P. Thus, P must decrease when rates increase.

f. When interest rates are at or below an implicit strike price (the strike price can be found by setting P = $30.6 million and solving for the 2-year yield) the bank must pay $30.6 million to Gibson. When rates rise above the implicit strike price, the bank saves by having to pay Gibson less. In other words, when interest rates rise above the implicit strike price, the bank has the right to "sell" Gibson back smaller and smaller amounts of principal.

CHAPTER 9 -- FINANCIAL CONTRACTING

A1. A **principal-agent relationship** exists whenever one party (the agent) makes decisions which affect the other party (the principal). Explicit examples of principal-agent relationships include the lawyer-client and money manager-investor relationships. Implicit examples of principal-agent relationships include the stockholder-manager, debtholder-stockholder, and employer-employee relationships.

A2. The **asset substitution problem** occurs when a firm with debt outstanding undertakes relatively high-risk investments even though such investments may reduce the overall market value of the firm. Because the debtholders are not compensated for the additional risk, the stockholders gain at the expense of the debtholders.

A3. The **underinvestment problem** occurs when a firm with risky debt outstanding has the incentive to forego an investment that would increase the total market value of the firm. When a positive-NPV investment decreases the risk of the firm, the debtholders share in the increased total market value.

A4. **Moral hazard** exits when the agent can take unobserved actions in the agent's own interest, to the detriment of the principal.

A5. A **free rider** is someone who receives the benefit of someone else's expenditure of money, effort, or creativity simply by imitation. The ability of people to free ride on the efforts of another person reduces that player's incentive to expend the effort in the first place. An example of the free rider problem is when others are able to copy and benefit from an inventor's valuable idea. Patent laws are the contract form that is typically used to reduce or eliminate this problem.

A6. **Agency problems** occur when someone (an agent) is hired to act in the interests of another (the principal) and the agent can take actions in his own interest at the expense of the principal. Three examples of the agency problem are the asset substitution problem, the underinvestment problem, and the problem of claim dilution.

A7. Three goals that managers might have that are not necessarily consistent with the goal of maximizing shareholder wealth are (1) maximizing growth, (2) maximizing perquisite consumption, and (3) diversifying their specific human capital.

A8. An **agency cost** is any cost involved in resolving principal-agent conflicts of financial self-interest. Agency costs are the incremental costs above what would be incurred in a perfect capital and labor market environment. The three components of agency costs are (1) financial contracting costs -- for example, incentive fees, (2) monitoring costs -- for example, auditing costs, and (3) misbehavior costs, that is, the loss of wealth the principal suffers as a result of the agent's pursuing divergent goals within an imperfect contract -- for example, the cost of elaborate offices. Agency costs are borne entirely by the principal.

A9. Asset uniqueness creates agency costs for shareholders in two ways: (1) There is greater risk associated with the disposal of unique assets so they have lower collateral value. This causes the debtholders to require a higher interest rate to compensate them for the increased risk. (2) When the firm's assets are unique, its employees' human capital will be even less well-diversified. This causes the employees to require higher pay.

A10. Employee perquisites create a conflict between the shareholders and the employees because they provide personal benefit to the employees but reduce the stockholders' residual claim on the earnings of the firm.

A11. Product and service guarantees can create an agency problem between the firm and its consumers because the consumers will only pay the firm "full value" for its products and services if they believe that the firm will fulfill its promise of future service.

A12. Four devices that naturally monitor agent behavior for the principal are: (1) The shareholder's right to elect the directors -- the threat that stockholders can elect new directors if they are dissatisfied with the firm's performance acts to monitor manager behavior; (2) Incentive compensation plans -- if the manager's compensation depends on the firm's performance, the manager's goals will be more closely aligned with the goals of shareholders; (3) The shareholder's right to sell their shares -- the ability to sell shares to prospective acquirers in a takeover attempt acts to monitor managers' actions; and (4) Competition in the managerial labor market -- the manager's desire to avoid being fired and to improve her future prospects causes the manager to act in ways that improve the firm's performance.

B1. Monitoring is not always the best choice because minimizing agency costs involves a trade-off of the three types of agency costs. For example, the costs of monitoring may exceed the costs due to occasional misbehavior on the part of the agent.

B2. When the manager is also a shareholder in the firm, her goals are more likely to be consistent with the goals of the other shareholders. This reduces the potential for manager actions that reduce shareholder value.

B3. **Claim dilution** due to a substantial increase in the amount of debt outstanding can be explained using the notion of stock as a call option on the firm's assets as follows: a substantial increase in the amount of debt outstanding increases the variability of the firm's cash flows and any increase in variability increases the value of the option; thus, the firm's stockholders benefit at the expense of the debtholders.

B4. **Human capital** consists of the unique capabilities and expertise that belong to an individual. A person's human capital is largely invested in a particular firm or industry, and these skills are not likely to be perfectly transferable to another job. As a result, human capital is rarely well-diversified; instead, it is concentrated in one area or profession.

B5. The stockholders' decision rule for choosing investments should be to undertake all positive-NPV investments. Due to the firm's expertise in a particular area, these investments are likely to be highly related to the firm's existing operations. This is not a concern to the stockholders because their investment in the firm is diversified by their other holdings. In contrast, because the manager has nondiversifiable human capital invested in the firm, she has the incentive to choose investments that diversify the firm's operations even if these investments are not as profitable for the firm.

B6. Because there is value in having a good reputation, a manager will be less likely to engage in activities that, if he is caught, will damage his reputation. As a result, the shareholders may be able to attain an acceptable level of manager compliance with shareholder objectives using less frequent or less stringent monitoring of the manager's activities than would be required if there were no value in having a good reputation.

B7. The stock price of a firm undergoing bankruptcy proceedings is usually positive and never negative because of the limited liability feature of common stock. The view of stock as an option on the assets of the firm illustrates this point. The worst that the stockholders can do is lose the entire value of their investment in a liquidation. If there is any possibility of the firm's shareholders recovering anything in a reorganization, then there is a positive value to the stockholder's option.

B8. Without bond covenants, shareholders may have the ability as well as the incentive to expropriate debtholder wealth through asset substitution, underinvestment, or claim dilution. Because rational debtholders expect the shareholders to act in their own self-interest, they will require a higher interest rate on bonds without covenants than they would otherwise accept on identical bonds with covenants.

B9. If undertaking a low-risk investment reduces the risk of the firm, then it reduces the variability of the return on the firm's assets. A reduction in variability reduces the value of a call option on those assets, thus reducing the value of the firm's shares. In spite of this share value reduction, the firm will increase in value by the NPV of the investment. If the shareholders' share of the positive NPV is smaller than the share value reduction due to the decrease in firm risk, shareholders will not want the firm to undertake the investment. If the investment is passed-up, the firm's debtholders are the losers.

B10. Debtholders typically require covenants that restrict the firm's ability to take on additional debt because a substantial increase in the leverage of the firm decreases the value of the firm's outstanding debt through claim dilution. Two common covenants related to this issue are restrictions against issuing debt that is of the same seniority as existing debt and restrictions against pledging assets for the benefit of other lenders. As long as the existing debt has a higher seniority than the newly issued debt, it will have a priority claim and will be less subject to the claim dilution associated with a new debt issue of the same seniority. Pledging assets for the benefit of other lenders gives those other lenders a priority claim on the pledged assets, thus diluting the claim of the existing debtholders.

B11. An optimal contract is one that balances the three types of agency costs so that the total cost is minimized. High contracting costs or high monitoring costs should result in lower costs of misbehavior, and low contracting and monitoring costs may result in higher costs of misbehavior, so there is a tradeoff of costs involved. The optimal contract achieves the tradeoff that minimizes the total cost.

B12. Financial distress intensifies conflicts of interest among the firm's claimants by distorting the agent's incentives. Recall that common stock can be viewed as a call option on the firm's assets. In this context, consider the common stock of a financially distressed firm. As it becomes more likely that the firm will not continue to exist in its present form, the agent's incentive may be to "get as much as possible out of the firm while it lasts" or to ensure that the firm continues to exist "at all cost." Asset substitution is particularly likely under financial distress because a high-risk investment may have the potential for a very high return needed to keep the firm alive for the shareholders. We saw this in the Green Canyon Project example. Managers are even more predisposed to choose investments that diversify the firm's operations to reduce the likelihood of losing their jobs even if these investments have a negative NPV. And, financial distress greatly reduces the value of a firm's guarantees of product quality and future service since there is a greater likelihood that the firm won't be around to uphold those guarantees.

B13. Shareholders can act to align managers' goals more closely with their own goals by providing appropriate financial incentives such as bonus and stock option plans, and by monitoring managerial actions and imposing penalties for failure to act in the shareholders' interest. They also have natural bonding devices such as the right to vote for directors who hire and fire managers and the right to sell their shares to bidders in a takeover attempt.

B14. Because a high-risk investment increases the variability of returns on the firm's assets, it increases the value of a call option on those assets. The shareholders gain more in positive outcomes and don't lose any more in negative outcomes with the high-risk investment. Thus, even if the investment has a negative NPV, the loss in total firm value comes entirely at the expense of the firm's debtholders while the shareholders' claim is increased in value.

C1. The question's own parenthetical remarks raise the issue of the trade-off between lower interest costs and reduced decision-making flexibility and the issue of the cost of overlapping restrictions versus their added protection. It should also be noted that a firm's ability to meet its debt payments is ultimately the result of its investment and operating decisions and not its bond covenants. While bond covenants serve important monitoring roles, they do not change the basic earning power of the firm and will not reduce the firm's cost of debt below some market-determined level. As the firm's cost of debt approaches this minimum, adding more covenants will create additional costs without offsetting benefits.

C2. It is not possible to have an agency problem without asymmetric information. If the principal had perfect information about the decisions and actions of the agent, then the agent would never be able to take actions that weren't in the best interest of the principal. If he did, the principal could impose penalties that effectively transfer the total costs of such actions back to the agent. These penalties would cancel out the benefits the agent received from acting in his interest at the expense of the principal, so he would have no incentive to pursue interests other than those of the principal.

C3. By allowing shareholders to get something in a bankruptcy proceeding, the firm's bondholders are reducing the possibility of future shareholder lawsuits or other similar actions which could delay the settlement of the bondholders' own claims. In effect, the bondholders would be offering a financial incentive to preempt potential legal action from the stockholders.

C4. As a debtholder, I would definitely be concerned about the firm's dividend policy. If the firm pays large dividends that reduce the value of the firm's assets and impair the firm's liquidity and future earnings, then, as a debtholder, I would share in the decreased value of the firm even though only the stockholders receive the dividend. In the extreme case, a large dividend could eliminate the firm's net worth. (As a result, state laws typically prevent a firm from declaring a dividend so large that it would render the firm insolvent.)

C5. Suppose the firm is in financial distress and it is in the best interests of the debtholders to liquidate the firm and pay off the firm's debts. If the managers choose to continue to operate the firm, simply because a large debt payment has not forced the firm into insolvency, there can be a loss to the firm's debtholders. In the case of an otherwise healthy firm, if excessive perquisite consumption were to impair the firm's ability to meet its debt obligations (interest and principal payments), then managers acting in their own interest would be doing so at the expense of *both* the debt- and equityholders. These possibilities create an agency problem between managers and the debtholders.

C6. A complex multilevel organization provides a natural form of agent monitoring by providing a system of checks and balances. For example, excessive perquisite consumption by certain levels of management is more difficult when other levels of management are monitoring that activity.

C7. Convertible debt helps to reduce the agency problem between the shareholders and the debtholders by giving debtholders the option of share ownership. A convertible bond can be modeled as a straight bond plus a call option on the stock of the firm, where the option can be exercised with a strike price equal to the value of the security as debt. This keeps the goals of the shareholders more in line with those of the convertible debtholders because if the shareholders take actions to benefit themselves at the expense of the debtholders, the debtholders can convert their bonds into stock and share in that benefit.

This point might be seen more easily in an alternative way. Recall our discussing put-call parity in Section 8.2 where we pointed out that any situation with a call option has a parallel description using a put option. It turns out that a convertible bond can alternatively be modeled as stock plus a put option. In that case, you can see that the convertible debtholders are, in effect, already owners. The difference is that the convertible debtholders also have, in addition to their ownership, a put option to sell their ownership back to the firm for its cash value at maturity. Thus, the firm cannot take actions to benefit shareholders to the exclusion of the convertible debtholders since the convertible debtholders are, in effect, already shareholders.

C8. Claimant coalitions would be expected to form primarily according to the priority of their claims. Claimants with the highest priority (senior debtholders and other debtholders) will always prefer liquidation if that is the positive-NPV decision. Lower priority claimants and those with an undiversified interest in the continuing operation of the firm (the common shareholders and managers) will prefer reorganizations and efforts that prolong operation because, with limited liability, the risk of the reorganization is borne by the higher priority claimants.

CHAPTER 10 -- COST OF CAPITAL

Note: Unless otherwise specified, assume that the tax rate is zero in the problems which follow.

A1. The firm's **financing decision** concerns the firm's choice between liabilities and owners' equity. The firm's **investment decision** concerns the firm's choice of assets in which to invest. The entities on the "other side" of these decisions are the suppliers of capital and the suppliers of real assets, respectively.

A2. A project that has an expected return exactly equal to its required return has zero NPV.

A3. We calculate Exxon's WACC using Equation (10.1).

$$WACC = (1 - 0.35)\times 0.14 + 0.35\times(1 - 0.40)\times 0.08 = 0.091 + 0.0168 = 10.78\%$$

A4. (a) First, use Equation (7.3) to determine the required return for each project:

$$r_j = r_f + \beta_j\times(r_m - r_f) = 9 + \beta_j\times(6)$$

Then, because the cash flows are perpetuities, NPV is computed as: NPV = CF/r_j - COST.

Project	beta	r_j	CF	NPV
A	1.00	15.00%	$310	$ 66.67
B	2.25	22.50	500	222.22
C	2.22	22.32	435	-51.08
D	0.65	12.90	270	93.02
E	1.37	17.22	385	235.77
F	2.36	23.16	450	-56.99

(b) The firm should undertake all positive-NPV projects, so it should undertake projects A, B, D, and E.

A5. (a) Using Equation (10.2), we can calculate β_A for the firm:

$$\beta_A = 0.20\times(1.00) + 0.10\times(2.25) + 0.12\times(2.22) + 0.10\times(0.65) + 0.34\times(1.37) + 0.14\times(2.36) = 1.5526$$

(b) Assume that the conglomerate's debt is riskless. (See Section [10.10] for a discussion of this.) Therefore, the beta of its debt is zero, and Equation (10.5) reduces to:

$$\beta_A = (1 - L)\times \beta$$

Rearranging the latter expression, we find beta for the equity.

$$\beta = \beta_A/(1 - L) = 1.5526/(1 - 0.20) = 1.941$$

A6. Present value depends on expected future cash flows and the required return.

A7. We approximate the beta with the same method used in A5. above.

$$\beta_A = (1 - L)\times\beta = (1 - 0.43)\times(1.32) = 0.752$$

A8. Leverage is the degree of total shareholder risk. **Operating leverage** is the degree of risk due to the firm's use of fixed versus variable cost production methods, and **financial leverage** is the degree of risk due to the firm's choice of fixed (that is, debt) versus variable cost (that is, equity) financing alternatives.

A9. If you own an asset with a required return equal to 10% and it has expected future cash flows of \$100 per year forever, then the present value of the investment is \$1000 (= 100/0.1). Suppose you have the opportunity to exchange your asset for an asset that would double your expected future cash flows to \$200 per year and the required return on that asset is 20%. The present value of the new asset is also \$1000 (= 200/0.2) so the increases in expected cash flows is perfectly offset by the increase in risk.

A10. Because the set of projects with expected return greater than the average cost of capital is riskier than the set of projects with expected return less than the average cost of capital, accepting only positive-NPV projects based on the average cost of capital requires accepting a set of projects that is riskier than the average project. Repeatedly accepting the riskier subset of potential projects will increase the risk of the firm and, also, the average cost of capital over time.

B1. We can compute the required return for each stock by applying the Capital Asset Pricing Model (CAPM):

$$r_j = r_f + \beta_j\times(r_m - r_f) = 9 + \beta_j\times(14 - 9) = 9 + \beta_j\times(5)$$

We can also compute the expected return to equity, r_e, using the dividend growth model, Equation (5.6), noting that the next expected dividend is equal to the last dividend times $(1 + g)$:

$$r_e = [D_0\times(1 + g)/P_0] + g$$

The results of these calculations are summarized below:

Stock	r_j	r_e
A	15.5%	13.4%
B	13.5	15.1
C	14.5	16.4

Invest in stocks B and C because in each case the expected return, r_e, exceeds the required return r_j.

B2. Operating risk depends on the nondiversifiable risk of the individual investment project and is reflected in the asset beta and required return for the project. Financial risk depends on the firm's financial leverage rather than on individual project characteristics. The firm can control its financial risk more readily than it can control its operating risk.

B3. We can estimate O'ryan's WACC with Equation (10.1), but first we must estimate the cost of debt, the cost of equity, and the proportions of debt and equity.

The before-tax cost of debt can be estimated as the YTM on the bonds. This is done with your calculator by entering the FV = 1000, PMT = 85, n = 21, and PV = 835 and computing the discount rate. This approximation assumes annual coupon payments and gives r_d = 10.47%.

The cost of equity is estimated as above in B1 by using the dividend growth model:

$$r_e = D_1/P_0 + g = (1.92/23.63) + 0.08 = 16.13\%$$

The total market value of the debt outstanding is \$83.5 million (100,000 bonds at \$835) and the total market value of the equity outstanding is \$236.3 million (10 million shares at \$23.63), so we estimate

$$L = 83.5/(83.5 + 236.3) = 0.26$$

Finally, applying Equation (10.1),

$$\text{WACC} = (1 - L)\times r_e + L\times(1 - T)\times r_d = (1 - 0.26)\times(16.13) + 0.26\times(1 - 0.34)\times(10.47) = 13.73\%$$

B4. In a perfect capital market environment, this purely financial transaction would be a zero-NPV transaction. The WACC for the firm is irrelevant to the decision because the required return depends only on the risk of the investment -- in this case, Treasury notes.

B5. a. Compute the YTM by finding the discount rate that solves the following equation:

$$295.49 = 200\times(1 + r)^8,$$

so that $r = 5\%$ per half year and YTM = 10%, compounded semiannually

b.

Period	Accrued Interest	Balance Due
1	\$10.00	\$210.00
2	10.50	220.50
3	11.03	231.53
4	11.58	243.11
5	12.16	255.27
6	12.76	268.03
7	13.40	281.43
8	14.07	295.50

B6. Because the project has the same risk as the market, the project beta is 1.0, and the required return on the project is 15%. With an expected return of 20%, the project has a positive NPV and should be undertaken even though the required return for the firm's current operations is much higher.

B7. First note that WACC = $(1 - L)\times r_e + L\times(1 - T)\times r_d = 0.20\times(25) + 0.80\times(0.60)\times(10) = 9.8\%$ and $r_f\times(1 - T) = 8\times(0.6) = 4.8\%$. Then WACC, r_e, and $(1 - T)\times r_d$ are graphed in Figure SM10-1 for values of L between 0 and 1. It is assumed in Figure SM10-1 that the value of the investment project is unaffected by how the project is financed.

Figure SM 10-1

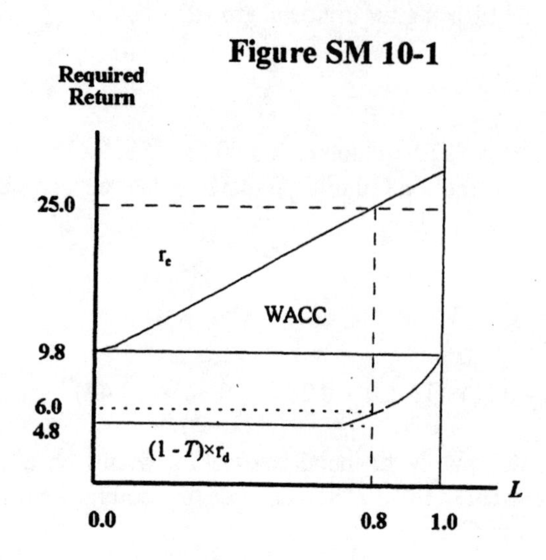

B8. Assuming an efficient capital market, the cost and present value of the investment are identical. Therefore, because you borrow \$6000 of a \$10,000 investment, L for the investment is 0.60. From this we can compute the required return for the investment:

$$\text{WACC} = (1 - L)\times r_e + L\times(1 - T)\times r_d = 0.4\times(20\%) + 0.6\times(1 - 0)\times(12\%) = 15.2\%$$

To find the return without leverage, solve the same equation with WACC constant and $L = 0$:

$$r_e = \text{WACC} = 15.2\%$$

B9. Case 1: For a project that is essentially identical to the firm with respect to operating risk and financing mix, the firm's WACC can be used as the required return. Alternatively, the firm's beta can be used to estimate the project beta and the CAPM used to estimate the required return. The proportion of debt financing is estimated from the market value of the firm's equity and book value of all the firm's liabilities; the equity beta is estimated by regression or found in an investor's guide; and the firm's beta is computed as the proportion of equity financing multiplied by the equity beta. Finally, the CAPM is applied using the firm beta to compute the required return for the project.

Case 2. For a project with risk significantly different from the average risk of the firm, the project beta must be estimated from outside sources. First, the firm identifies a set of firms whose primary business is closely related to the potential project and estimates equity betas for the firms in that set. Next, the proportion of debt in each of these firms is estimated from balance sheet data and equity market values, and firm betas are estimated for each firm in the set using the method in Case 1. Finally, these firm betas are averaged to approximate the project beta and the CAPM is applied to compute a required return for the project.

B10. First, find the semi-monthly interest rate that solves the following equation:

$$20{,}000 = \sum_{t=1}^{120} \frac{300.99}{(1+r)^t}$$

Using your calculator with PV = 20,000; PMT = 300.99; $n = 5\times(12)\times(2) = 120$, FV = 0; compute r = 1.10% per half-month. This corresponds to an APR = 1.1×(24) = 26.4% and APY = $(1.011)^{24} - 1$ = 30.0%.

B11. Because A and B are mutually exclusive projects with equal lives, the appropriate decision rule is to choose the project with the highest NPV. To compute the NPV for each project, we need to know the required return for each and, therefore, we first need to know the beta for each project:

$$\beta_A = \frac{Corr(A,M)\times\sigma_A\times\sigma_M}{\sigma_M^2} = \frac{0.8\times\sqrt{0.25}\times\sqrt{0.16}}{0.16} = 1.0$$

$$\beta_B = \frac{Corr(B,M)\times\sigma_B\times\sigma_M}{\sigma_M^2} = \frac{0.32\times\sqrt{0.25}\times\sqrt{0.16}}{0.16} = 0.4$$

Then, applying the CAPM, we get the required return for each project:

$$r_A = 9\% + 1.0\times(14\% - 9\%) = 14\%$$

$$r_B = 9\% + 0.4\times(14\% - 9\%) = 11\%$$

$$NPV_A = 114{,}000/(1.14) - 100{,}000 = 0 \text{ (Required return = Expected return)}$$

$$NPV_B = 56{,}000/(1.11) - 50{,}000 = 450.45 \text{ (Expected return > Required return)}$$

Therefore, the firm should choose project B.

C1. You could make a profit from this situation by purchasing the firm and liquidating it: Buy all of the firm's outstanding debt and equity for $8.5 million and receive $10 million from the sale of the assets, leaving a riskless profit of $1.5 million. Alternatively, you could short sell the assets of the firm for $10 million and use the proceeds to buy all the outstanding debt and equity in the firm. You would have $1.5 million left over and because you bought all the debt and equity in the firm, you now own the assets of the firm and can cover your short sale. The result is the same, a riskless profit of $1.5 million.

C2. a. We will use the average of the unleveraged betas for the four oil companies as an estimate of the beta for the project. To do this, we must apply each firm's L to compute the unleveraged beta using Equation (10.5):

Company	β	L	β_A
Franklin Oil	1.40	0.50	0.8434
Oscar Oil	1.45	0.50	0.8735
VMB Oil & Gas	1.30	0.45	0.8442
Peters Oil & Gas	1.55	0.60	0.7789

Averaging the four values for oil and gas companies gives $\beta_A = 0.8350$

b. The required return on unleveraged equity is

$$r_e = WACC = r_f + \beta_A\times(r_m - r_f) = 10 + 0.8350\times(6) = 15.01$$

c. To find the required return on the levered equity, we need to find the levered beta by rearranging Equation (10.5):

$$\beta = \beta_A \times (1 - T \times L)/(1 - L) = 0.8350 \times (1 - 0.34 \times .5)/(1 - 0.5) = 1.40$$

$$r_e = r_f + \beta \times (r_m - r_f) = 10 + 1.40 \times (6) = 18.4\%$$

Finally, the weighted average cost of capital is computed from Equation (10.4):

$$\text{WACC} = (1 - 0.5) \times (18.4) + (0.5) \times (1 - 0.34) \times (15.15) = 14.2\%$$

C3. a. Following the same procedure as above in C2, the unleveraged beta for each firm is computed and the average provides an estimate of the unleveraged beta for the project:

Firm	β_A
A	0.9016
B	0.9354
C	0.9871
D	0.7831
E	1.0246

The average gives $\beta_A = 0.9326$ and releveraging gives $\beta = 0.9326 \times (1 - 0.34 \times 0.33)/(1 - 0.33) = 1.236$.

b. The required return on the project equity is computed by applying the CAPM:

$$r_e = 9 + 1.236(6) = 16.42\%$$

c. $\text{WACC} = (1 - 0.33) \times (16.42) + (0.33) \times (1 - 0.34) \times (11) = 13.40\%$

C4. Using Equation (10.4) with a tax rate of 34% gives

$$(1 - TL) \times \beta_A = L \times \beta_d + (1 - L) \times \beta$$

$$(1 - (0.34) \times (0.72) \times \beta_A = (0.7552) \times \beta_A = 0.72 \times (0.32) + (1 - 0.72) \times (3.1) = 1.10$$

So $\beta_A = 1.10/(0.7552) = 1.46$

C5. The CFO should explain to the chairman that his thinking is incorrect. In a perfect capital market environment, the value of the firm does not depend on its capital structure. This means that as long as the business risk of the firm is unchanged, the required return to the firm's assets will remain constant. In such a setting, the cost of equity will always be above the cost of debt and both will increase with increases in the proportion of debt outstanding. The required return on each security perfectly reflects the risk-return trade-off of that security. If the firm issues "cheap" debt and retires "expensive" equity, both rates increase but the average remains the same. (See Figure 10-6 for an illustration of this phenomenon.) The chairman is missing the fact that equity's risk per dollar invested changes as L changes. (The chairman is "falling for" the same misleading outcome illustrated in the Per-Pet example on page 316-317.)

C6. The value, at $t = 5$, of the five $7072.12 payments you make must equal the value, at $t = 5$, of the $200 monthly perpetuity you will receive. The value (at $t = 5$) of the perpetuity you will receive is the perpetuity divided by the monthly rate, or $\$200/r_m$ and the value (at $t = 5$) of the annuity you will pay is given by

$$FVA_5 = \$7072.12 \times \left[\frac{(1 + APY)^5 - 1}{APY}\right]$$

where $APY = (1 + r_m)^{12} - 1$.

Setting the two equations equal, choosing possible values for r_m, and solving by trial and error gives APY = 6.17%:

r_m	APY	PV of perpetuity	FV of annuity
1.00%	12.68%	$20,000	$45,539
0.75	9.38	26,667	42,646
0.50	6.17	40,000	40,000

CHAPTER 11 -- CAPITAL BUDGETING: THE BASICS

A1. The four basic steps involved in evaluating an investment for an individual or a corporation are:
1. Estimate the expected future cash flows;
2. Assess the risk and determine a required return (cost of capital) for discounting the expected future cash flows;
3. Compute the present value of the expected future cash flows; and
4. Compare the investment's cost to it present value.

A2. The Caltron Company has the following before-tax cash flows associated with the alternatives of **expensing** or **capitalizing** the asset it just purchased:

Time	0	1	2	3	4	5
Expensed	0.5M	0	0	0	0	0
Capitalized	0	0.1M	0.1M	0.1M	0.1M	0.1M

For any required return greater than zero, the tax benefit to Caltron of expensing the purchase of the asset will be larger than that from depreciating the asset.

A3. When we are analyzing a capital budgeting project, we want to know the difference between the firm's cash flows with and without the project. The incremental changes in the firm's cash flows which occur as a result of adopting the project are what matter, because these changes indicate whether the project's adoption adds or subtracts value. Applying the Principle of Incremental Benefits enables us to both identify those cash flows which are relevant to the decision and those which are irrelevant.

A4. Because evaluating a project requires identifying its relevant incremental cash flows and because taxes are paid in cash, any tax effects due to the current tax laws are extremely important to evaluating a project properly.

A5. **Sunk costs** are excluded from the valuation of a capital investment project because they have already been incurred and cannot be affected by the decision to accept or reject a project.

A6. Financing charges are normally accounted for in the required return or discount rate.

A7. By Equation(11.5), net salvage value (NSV) is: $NSV = (1 - T)\times S + T \times B - (1 - T)\times REX + \Delta W$; so that

$$NSV = (1 - 0.34)\times 18{,}000 + 0.34 \times (10{,}000) - (1 - 0.34)\times 1000 + 0 = \$14{,}620$$

A8. The internal rate of return is the project's expected return, the discount rate that would make the project's NPV = 0.

A9. Investments are **mutually exclusive** if acceptance of one precludes acceptance of any of the others.

A10. The **profitability index** is the present value of all the project's expected cash flows divided by the initial outlay. This is the same as one plus the project's NPV divided by the initial outlay.

A11. The **payback** of a project is the time it takes the firm to recover its initial cash investment in the project.

A12. The Principle of Valuable Ideas is of critical importance to the capital budgeting process because it suggests places to look for potential investment project opportunities, and generating ideas for capital budgeting projects is the critical first step in the capital budgeting process.

A13. NPV is the most reliable investment criterion because it incorporates all the investment's cash flows, assumes a reasonable discount/reinvestment rate, and is a measure of the value that the acceptance of the investment would imply. In contrast, both regular and discounted payback ignore cash flows which occur after the payback period, and IRR assumes that cash flows are reinvested at the IRR.

A14. The NPV profile includes both the NPV and IRR, as well as displaying the value of the project at different costs of capital. It provides the most complete view of the project and enables one to identify costs of capital at which the project would and would not add value.

B1. a. To compute the NPV of the replacement investment, we need to compute the incremental cash flows. First, we compute the net initial outlay using Equation (11.2):

$$\begin{aligned} C_0 &= -I_0 - \Delta W - (1 - T)\times E_0 + (1 - T)\times S_0 + T\times B_0 + I_c \\ &= -300{,}000 - 0 - (1 - 0.30)\times 30{,}000 + (1 - 0.30)\times 25{,}000 + 0.30 \times 60{,}000 + 0 \\ &= -285{,}500 \end{aligned}$$

The new machine produces \$80,000 per year in pretax operating savings, so $\Delta E = -\$80{,}000$, but does not increase revenues. Depreciation on the new machine is \$41,667 per year (= (300,000 - 50,000)/6), but in years 1-3 \$20,000 per year (= 60,000/3) is lost as a result of the sale of the old machine. So, from Equation (11.4):

$$\begin{aligned} CFAT_{1\text{-}3} &= (1 - T)\times(\Delta R - \Delta E - \Delta D) + \Delta D \\ &= (1 - 0.30)\times(0 - (-80{,}000) - 21{,}667) + 21{,}667 \\ &= 62{,}500, \text{ and} \end{aligned}$$

$$CFAT_{4\text{-}6} = (0.70)\times(80{,}000 - 41{,}667) + 41{,}667 = 68{,}500.$$

The net salvage value for the new machine is given as \$40,000.

b.

$$0 = -285{,}500 + \sum_{t=1}^{3} \frac{56{,}000 + 12{,}500 - 6000}{(1 + \text{IRR})^t} + \sum_{t=4}^{6} \frac{56{,}000 + 12{,}500}{(1 + \text{IRR})^t} + \frac{40{,}000}{(1 + \text{IRR})^6}$$

The IRR for the project is about 12.23%.

c. From Equation (11.6), at 12% the project's NPV is \$1986.

$$\text{NPV} = -285{,}500 + \sum_{t=1}^{3} \frac{56{,}000 + 12{,}500 - 6000}{(1.12)^t} + \sum_{t=4}^{6} \frac{56{,}000 + 12{,}500}{(1.12)^t} + \frac{40{,}000}{(1.12)^6} = \$1986$$

In by-item form:

Time	Item	CFBT	CFAT	PV@12%
0	Capitalized equipment cost	-300,000	-300,000	-300,000
0	Expensed equipment cost	-30,000	-21,000	-21,000
0	Sale of old equipment	+25,000	+35,500	35,500
1-3	Lost depreciation from sale of old equipment	0	-6,000	-14,411
1-6	Depreciation	0	+12,500	51,393
1-6	Change in expenses	+80,000	+56,000	230,239
6	Net salvage value	Not needed	+40,000	20,265
			NPV =	1,986

d. The project's NPVs at different costs of capital are:

Cost of capital	NPV
0%	$147,500
4%	88,549
8%	40,912
12%	1986
16%	-30,153

B2. To compute the NPV of the replacement investment, we need to compute the incremental cash flows. First, we compute the net initial outlay using Equation (11.2):

$$C_0 = -I_0 - \Delta W - (1 - T)\times E_0 + (1 - T)\times S_0 + T\times B_0 + I_c$$
$$= -65{,}000 - 0 - (1 - 0.4)\times 0 + (1 - 0.4)\times 30{,}000 + 0.4\times(50{,}000) + 0 = -27{,}000$$

Annual revenues from the old machine are $60,000 and annual revenues from the new machine are expected to be $70,000, so the change in revenues is ΔR = $10,000 (= 70,000 - 60,000). Similarly, annual operating expenses with the old machine are $30,000, and annual operating expenses with the new machine are expected to be $25,000, so ΔE = -$5000 (= 25,000 - 30,000). Because the old machine has a current book value of $50,000 and is being depreciated to zero over its remaining 5 years of useful life, annual depreciation on the old machine is $10,000 (= 50,000/5). Annual depreciation on the new machine will be $12,000 (= [65,000-5000]/5). The change in depreciation with the machine replacement will be ΔD = $2000 (= 12,000 - 10,000). With these values we can compute the net operating cash flows using Equation (11.3):

$$\text{CFAT(1 through 5)} = (1 - T)\times(\Delta R - \Delta E) + \Delta D = (1 - 0.4)\times[10{,}000 - (-5000)] + 0.4\times(2000) = 9800$$

There are no non-operating cash flows expected with this machine replacement, but there will be a salvage value associated with disposal of the new machine at the end of five years. Because the machine is expected to have a market value equal to its book value at the end of five years, there are no tax consequences associated with the sale at that time. With no removal and cleanup expenditure and with no recovery of net working capital, the net salvage value (from Equation (11.5)) will equal the salvage value, $5000.

The NPV from Equation (11.6) is the present value of all the incremental after-tax cash flows connected with the replacement:

$$\text{NPV} = -27{,}000 + \sum_{t=1}^{5} \frac{9800}{(1.10)^t} + \frac{5000}{(1.10)^5} = \$13{,}254.32$$

[On a calculator, (Step 1) PMT = 9800, n = 5, r = 10, FV = 5000; compute PV = 40,254.32 minus 27,000 = $13,254.32.]

B3. The firm has the following before-tax cash flows associated with the alternatives of **expensing** or **capitalizing** the asset it just purchased:

Time	0	1	2	3
Expensed	9000	0	0	0
Capitalized	0	3000	3000	3000

The PV of the tax shields of each of these alternatives is computed as follows:

$$\text{PV}_\text{E} = 9000 \times (0.38) = \$3420$$

$$\text{PV}_\text{C} = \sum_{t=1}^{3} \frac{3000 \times (0.38)}{(1.12)^t} = \$2738.09$$

The present value difference between the two alternatives is therefore $681.91 (= 3420 - 2738.09).

B4. a. The NPV for each project is found by discounting the annual after-tax cash flows and deducting the cost of the project:

$$\text{NPV}_\text{A} = -80{,}000 + \sum_{t=1}^{6} \frac{40{,}222.22}{(1.12)^t} = \$85{,}369.93$$

[Put in n = 6, r = 12, PMT = 40,222.22, FV = 0, then compute PV = 165,369.93 minus 80,000 = $85,369.93.]

$$\text{NPV}_\text{B} = -80{,}000 + \sum_{t=1}^{5} \frac{44{,}966.67}{(1.12)^t} = \$82{,}094.78$$

[Put in n = 5, r = 12, PMT = 44,966.67, FV = 0, then compute PV = 162,094.78 minus 80,000 = $82,094.78.]

b. The internal rate of return for each project is the discount rate that makes NPV = 0. Thus, set the NPVs in the above equations equal to zero and solve for the discount rate, r:

IRR_A = 44.83% and IRR_B = 48.40%

[Note that a zero NPV is the same as a PV equal to the cost. We can find the IRR by entering PMT = CFAT, PV = 80,000, n = 5, FV = 0, and solving for r.]

c. Solve for the crossover point (the point where the NPV of the two projects are equal) by trial and error by choosing alternative discount rates, and computing the two NPVs:

r	NPV_A	NPV_B
25%	38,713	40,928
20	53,759	54,478
19	57,149	57,492
18	60,681	60,618
18.2	59,963	59,984
18.1	60,321	60,301
18.15	60,142	60,142

Here is a graph of the NPV of the projects as a function of the discount rate:

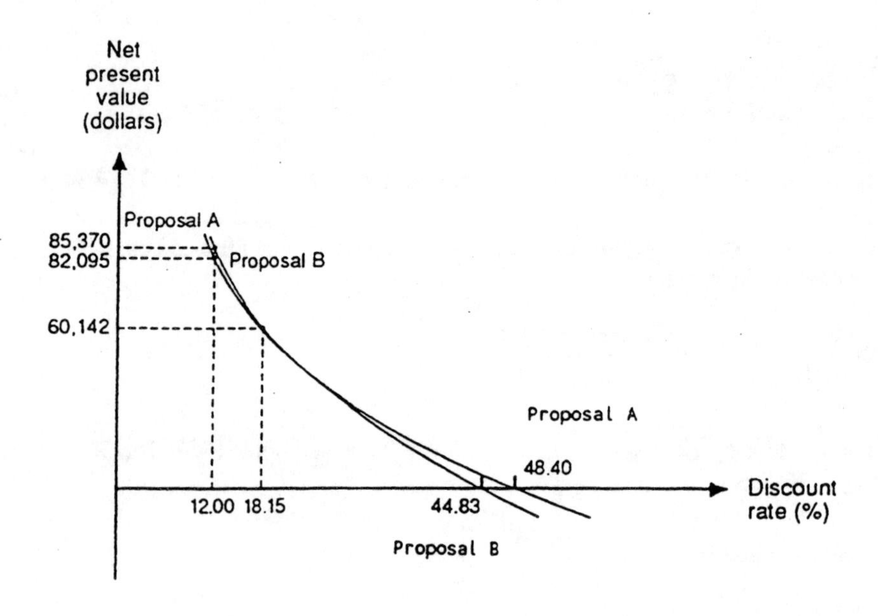

d. You should undertake project A because it has the higher NPV.

B5. The cash flow items associated with the proposed project are calculated below:

$C_0 = -I_0 = -2500$, and Net Salvage Value $= 0$

CFAT(1 through 5) $= (1 - T)\times(\Delta R - \Delta E) + \Delta D = (1 - 0.4)\times(300 - (-500)) + 0.4\times(500) = 680$

$$NPV = -2500 + \sum_{t=1}^{5} \frac{680}{(1.12)^t} = \$ -48.75$$

B6. a. Compute the NPV by discounting the expected cash flows:

$$NPV = -100 + \frac{25}{(1.1)} + \frac{50}{(1.1)^2} + \frac{50}{(1.1)^3} + \frac{25}{(1.1)^4} + \frac{10}{(1.1)^5} = \$24.90$$

The firm should accept the project because NPV > 0.

b. Setting the NPV equation in part a equal to zero and solving for the discount rate gives IRR = 20.28%. The firm should accept the project because IRR > hurdle rate.

[This problem can be solved directly if your calculator has uneven cash flow capabilities. Otherwise you must use trial and error to find the IRR.]

c.

$$0 = -100 + \sum_{t=1}^{PAYBACK} CFAT_t$$

$0 = -100 + 25 + 50 + 0.5 \times (50)$; therefore, payback $= 2.5$ years

d. Not very much. Payback provides some indication of the project's risk in that it estimates the time until the initial investment is recovered without regard to the time value of money.

e. If the project is one of two mutually exclusive projects and the NPV and IRR criteria lead to different choices, the firm should accept the project with the largest NPV.

B7. a. The firm should accept the project because NPV > 0:

$$NPV = 50 + \frac{100}{(1.12)} + \frac{-20}{(1.12)^2} + \frac{-50}{(1.12)^3} = \$87.75$$

b. Solving for the discount rate that makes NPV = 0 gives IRR = -31.12%. The *standard* IRR decision rule would reject the project.

c. The IRR is defined as the discount rate for which NPV = 0. Because of the nature of the cash flows in this example, that rate is a negative value.

d. The IRR decision rule is appropriate for evaluating economically independent projects for which all after-tax cash flows but the first are positive. In this case the cash flow pattern is reversed (so to speak): two inflows followed by two outflows. For situations where inflows are followed by outflows (with only one flow reversal) a modified IRR decision rule applies that is the reverse of the standard rule: accept the project when the IRR is *less* than the hurdle rate. Because the cash flow pattern has reversed, the firm is effectively "borrowing" 50 and 100 and then "repaying" 20 and 50. If it can "borrow" more cheaply than the required return, its shareholders benefit, which is the reason the IRR decision rule reverses. Also, the rate in this case is negative because the time value of money must cause the inflows in 0 and 1 to *decrease* in value into the future to be exactly worth the required outflows. That is, the total inflow of $150 (= 50 + 100) must decline in value down to $70 (= 20 + 50).

B8. When considering mutually exclusive projects, the NPV decision rule is the appropriate criterion to apply.

$$NPV_A = -100 + \frac{30}{(1.14)} + \frac{40}{(1.14)^2} + \frac{50}{(1.14)^3} + \frac{40}{(1.14)^4} + \frac{30}{(1.14)^5} = \$30.11$$

$$NPV_B = -150 + \frac{45}{(1.14)} + \frac{60}{(1.14)^2} + \frac{75}{(1.14)^3} + \frac{60}{(1.14)^4} + \frac{60}{(1.14)^5} = \$52.95$$

The firm should accept project B because its NPV is $52.95, while the NPV of A is $30.11. When considering mutually exclusive projects, the NPV decision rule is the correct rule to apply.

B9. a. Project A IRR = 20.86%; Project B IRR = 19%

b.

$$NPV_A = -350 + \frac{140}{(1.10)} + \frac{140}{(1.10)^2} + \frac{100}{(1.10)^3} + \frac{100}{(1.10)^4} + \frac{65}{(1.10)^5} + \frac{30}{(1.10)^6} = \$93.70$$

$$NPV_B = -350 + \frac{65}{(1.10)} + \frac{65}{(1.10)^2} + \frac{100}{(1.10)^3} + \frac{140}{(1.10)^4} + \frac{140}{(1.10)^5} + \frac{175}{(1.10)^6} = \$119.28$$

c.

Cost of Capital	Project A NPV	Project B NPV
14%	54.66	59.86
18%	21.16	10.86
22%	-7.83	-29.91

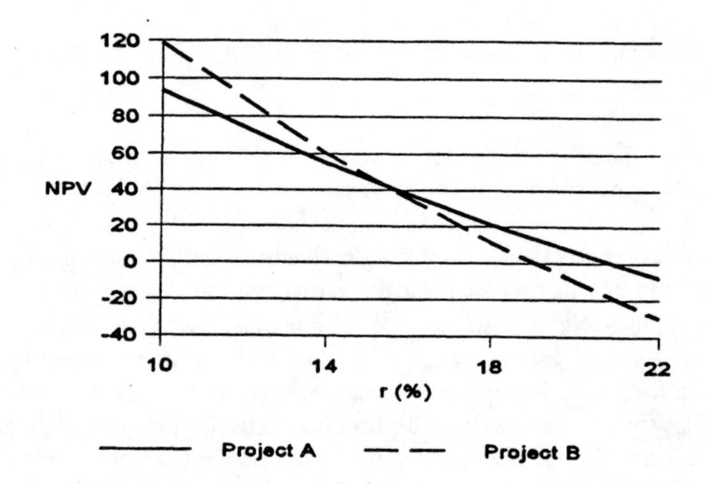

d. Based on viewing the NPV Profile, at a 12% cost of capital Kodak should undertake project B. Project B has the higher NPV (88.10 versus 73.42), and the NPV rule should be used when comparing mutually exclusive projects.

B10. a. The NPVs and IRRs are computed as follows:

$$NPV_A = -100 + \frac{25}{(1.15)} + \frac{30}{(1.15)^2} + \frac{40}{(1.15)^3} + \frac{30}{(1.15)^4} + \frac{25}{(1.15)^5} = \$0.31$$

$$NPV_B = -50 + \frac{10}{(1.15)} + \frac{15}{(1.15)^2} + \frac{25}{(1.15)^3} + \frac{15}{(1.15)^4} + \frac{15}{(1.15)^5} = \$2.51$$

$IRR_A = 15.13\%$; $IRR_B = 16.97\%$

b. If the projects are independent, then both should be accepted because both have a positive NPV and, an IRR that is greater than the hurdle rate of 15%.

c. Assuming dependence means that accepting one project requires the acceptance of the other, and assuming the cash flows for each are unaffected by the dependence, the NPV is additive, and both projects should be accepted.

d. If the projects are mutually exclusive, project B should be chosen because it has the higher NPV.

B11. Fuji's expected cash flow pattern is:

Time	0	1	2	3	4	5
Cash flows	-500	100	150	150	200	200
PV	-500	89	120	107	127	113
Cumulative	-500	-411	-291	-184	-57	56

Discounted payback = 4.5 years (= 4 + 57/113).

B12. The discounted payback is the amount of time it takes to recover the project's initial cost. Thus, discounted payback for a conventional project is the amount of time it will take for the project's NPV to equal zero.

B13. The cash flows and NPV associated with the TX2 are:

Time	Item	CFBT	CFAT	PV @ 15%
0	Capitalized cost	-60,000	-60,000	-60,000
0	Expensed setup cost	-10,000	-6000	-6000
0	Change in net working capital	-10,000	-10,000	-10,000
1-6	Depreciation	0	4000	15,138
1-10	Change in revenues minus expenses	17,500	10,500	52,697
7	Maintenance cost	-30,000	-18,000	-6767
10	Sale of equipment	20,000	12,000	2966
10	Return of net working capital	10,000	10,000	2472
			NPV =	-$9494

B14. Because any increase in working capital ties up money that would be available for other uses and because there is a time value of money, it is important to account for the opportunity cost of the unavailable funds when evaluating a potential investment project.

B15. The cash flows and NPV associated with the Miller project are:

Time	Item	CFBT	CFAT	PV @ 16%
0	Capitalized cost	-6,000,000	-6,000,000	-6,000,000
0	Change in net working capital	-500,000	-500,000	-500,000
1-6	Depreciation	0	262,500	967,243
1-2	Change in revenues minus expenses	5,000,000	3,250,000	5,217,004
3-6	Change in revenues minus expenses	3,000,000	1,950,000	4,055,033
6	Sale of equipment	0	525,000	215,482
6	Return of net working capital	500,000	500,000	205,221
			NPV =	$4,159,983

Note that computing the PV of the 4-year, $1,950,000 annuity representing change in revenues minus expenses for years 3-6 gives the value at year 2 and this amount must be discounted as a lump sum for 2 years to get the time 0 present value reported in the table. Also note that, because the equipment cost is depreciated over eight years, the annual depreciation expense is $750,000. However, the asset is sold at the end of 6 years so only 6 years of tax shield are realized, and the asset has a book value (cost - accumulated depreciation) of $1,500,000 at the time it is sold. This write-off for tax purposes accounts for the $525,000 cash flow upon the sale of the equipment even though the salvage value (net of removal costs) is zero.

C1. Payments to shareholders and to meet the firm's financial obligations come only from cash flow. Therefore, direct cash flows should be identified and measured first in the capital budgeting process in order to assure that those obligations are met. Once cash and quantified items have been accounted for, non-cash and non-quantified items can be considered. These non-cash or non-quantified items can be important, but are most easily and properly examined only after the numbers have been crunched.

C2. To compute the minimum amount that the government would have to pay Doug at time $t = 0$ to induce him to undertake the project, compute the present value of cash flows associated with the operation and then set the price so as to make the NPV zero. The appropriate discount rate for the project is:

$$r_j = r_f + \beta_j \times (r_M - r_f) = 5 + 1.3 \times (12 - 5) = 14.1\%$$

The cash flows for the project are summarized below (explanatory notes follow the table):

Time	Item	CFBT	CFAT	PV @ 14.1%
1	Land cost	-250,000	-250,000	-219,106
2	Capitalized building cost	-2,000,000	-2,000,000	-1,536,239
2	Capitalized equipment cost	-3,000,000	-3,000,000	-2,304,358
3	Investment tax credit	0	310,000	208,692
3	Change in net working capital	-100,000	-100,000	-67,320
4-33	Depreciation-building	0	26,000	121,763
4-33	Depreciation-equipment	0	240,000	1,123,963
4-33	Change in revenues minus expenses	550,000	330,000	1,545,550
8	Equipment replacement	-3,000,000	-3,000,000	-1,044,326
13	Equipment replacement	-3,000,000	-3,000,000	-540,018
18	Equipment replacement	-3,000,000	-3,000,000	-279,241
23	Equipment replacement	-3,000,000	-3,000,000	-144,395
28	Equipment replacement	-3,000,000	-3,000,000	-74,666
33	Sale of land	250,000	250,000	3217
33	Sale of building	50,000	50,0000	643
33	Restoration expense	-420,000	-252,000	-3243
33	Return of net working capital	100,000	100,000	1287
			Total PV =	-$3,207,797

For the items that are 30-year annuities over the period $t = 4\text{-}33$, computing the present value of the annuity gives the value at $t = 3$, and these amounts must be discounted as lump sums for 3 years to get the $t = 0$ present values reported in the table above.

The change in net working capital is computed as the change in current assets minus current liabilities:

20,000(cash) + 60,000(A/R) + 80,000(inventory) - 60,000(A/P)

The minimum amount the government would have to pay for the contract is the amount that would make the NPV zero, which is $3,207,797.

C3. The required return for computing the NPV is the rate that reflects the risk of the project, that is, the risk of the publishing business. We can use the information about the firm to find the riskless return and then apply that in the calculation of the required return for the project:

$r_e = 15\% = r_f + 2.0 \times (10 - r_f) = 20 - r_f$; so that $r_f = 5\%$; and, therefore,

$r^* = 5 + 1.4 \times (10 - 5) = 12\%$

Time	Item	CFBT	CFAT	PV @ 12%
0	Building	-80,000	-80,000	-80,000
0	Change in net working capital	-60,000	-60,000	-60,000
1-8	Depreciation	0	3200	15,896
1-2	Change in revenues minus expenses	30,000	18,000	30,421
3-8	Change in revenues minus expenses	50,000	30,000	98,328
6	Sale of equipment	50,000	36,400	14,701
6	Return of net working capital	60,000	60,000	24,233
			NPV =	$43,579

CHAPTER 12 – CAPITAL BUDGETING: SOME COMPLICATIONS

A1. **Erosion**occurs when the introduction of a product innovation decreases the profitability of one or more existing products, and **enhancement** occurs when the introduction of a product innovation increases the profitability of one or more existing products. Potential erosion and enhancement of existing products must be considered when evaluating a potential investment project.

A2. An **equivalent annual cost** is a constant amount that, if it were paid each year of a project's life, would have the same present value as the actual total costs of the project.

A3. Using Equation (12.2):

$$EAC = TC \times \left[\frac{r \times (1 + r)^n}{(1 + r)^n - 1}\right] = 73{,}285 \times \left[\frac{0.11 \times (1.11)^7}{(1.11)^7 - 1}\right] = \$15{,}552$$

A4. Under the optimistic sales level: CFBT = 5 ×(250,000) - 500,000 = 750,000

Time	Item	CFBT	CFAT	PV @ 12%
0	Salvage (forgone)	-1,200,000	-1,200,000	-1,200,000
1-6	ΔR - ΔE	750,000/yr	450,000/yr	1,850,133
1-4	Depreciation	0	60,000/yr	182,241
			NPV =	$832,374

Under the pessimistic sales level: CFBT = 5×(150,000) - 500,000 = 250,000

Time	Item	CFBT	CFAT	PV @ 12%
0	Salvage (forgone)	-1,200,000	-1,200,000	-1,200,000
1-6	ΔR - ΔE	250,000/yr	150,000/yr	616,711
1-4	Depreciation	0	60,000/yr	182,241
			NPV =	-$401,048

A5. a. The firm should undertake the project because it has a positive NPV:

$$NPV = \sum_{t=1}^{6} \frac{CF_t}{(1.13)^t} = \$19.91$$

b. To find the optimal abandonment time, sequentially determine (for each time period) whether the present value of the remaining cash flows exceeds the abandonment value:

At $t = 1$, the present value of the remaining cash flows is:

$$PV = \frac{30}{(1.13)} + \frac{30}{(1.13)^2} + \frac{20}{(1.13)^3} + \frac{20}{(1.13)^4} + \frac{10}{(1.13)^5} = 81.60$$

This is greater than the $40 abandonment value so the project would not be abandoned at $t = 1$. At $t = 2$, the present value of the remaining cash flows is:

$$PV = \frac{30}{(1.13)} + \frac{20}{(1.13)^2} + \frac{20}{(1.13)^3} + \frac{10}{(1.13)^4} = 62.21$$

This is greater than the $35 abandonment value so the project would not be abandoned at t = 2. At t = 3, the present value of the remaining cash flows is:

$$PV = \frac{20}{(1.13)} + \frac{20}{(1.13)^2} + \frac{10}{(1.13)^3} = 40.29$$

This is greater than the $30 abandonment value so the project would not be abandoned at t = 3. At t = 4, the present value of the remaining cash flows is:

$$PV = \frac{20}{(1.13)} + \frac{10}{(1.13)^2} = 25.53$$

This is greater than the $20 abandonment value so the project would not be abandoned at t = 4. At t = 5, the present value of the remaining cash flows is:

$$PV = \frac{10}{(1.13)} = 8.85$$

This is less than the $10 abandonment value so the project would be abandoned at t = 5.

A6. a. Using Equation (12.5),

EVPI = EV(perfect test) - EV(no test) = 1,010,000 - 734,500 = $275,500

b. Information value = EV(test) - EV(no test) + Test cost = 853,225 - 734,500 + 87,000 = $205,725

A7. **Simulation** is a technique that uses a mathematical model to simulate (in a sense, project) possible outcomes and to infer from these possible outcomes the best course of action to take (for example, whether to accept or reject a proposed investment project).

A8. **Sensitivity analysis** is the process of examining the sensitivity of the outcome to a possible decision (for example, undertaking an investment project) to particular parameters (for example, product price or the cost of a critical necessary asset).

A9. First, the incremental cash flows associated with the X-tender are computed. Applying Equation (11.2),

$$C_0 = -I_0 - \Delta W - (1 - T)\times E_0 = -120{,}000 - 20{,}000 - (1 - 0.45)\times 20{,}000 = -151{,}000$$

With a capitalized cost of $120,000 depreciated straight-line over 6 years to a zero salvage value, the change in depreciation will be $20,000 for years 1-6 and zero for years 7-10. Applying Equation (11.3),

CFAT(1 through 6) = $(1 - 0.45)\times 35{,}000 + 0.45\times(20{,}000) = 28{,}250$

CFAT(7 through 10) $= (1 - 0.45)\times 35{,}000 = 19{,}250$

The net salvage value (NSV) is found by applying Equation (11.5):

$$NSV = (1 - T)\times S + T \times B + \Delta W = (1 - 0.45)\times 40{,}000 + 0.45\times(0) + 20{,}000 = 42{,}000$$

Finally, the NPV is the present value of the incremental cash flows at the project's required return:

$$NPV = -151{,}000 + \sum_{t=1}^{6}\frac{28{,}250}{(1.16)^t} + \sum_{t=7}^{10}\frac{19{,}250}{(1.16)^t} + \frac{42{,}000}{(1.16)^{10}} = -\$15{,}277.04$$

A10. We compute the EAC under each scenario:

Number of years	4	5	6	7
Initial cost	-10,000	-10,000	-10,000	-10,000
Net salvage value	3800	2800	1000	-1000
PV of NSV at 12%	2415	1589	507	-452
Total cost	-7585	-8411	-9493	-10,452
EAC	-2497	-2333	-2309	-2290

The EAC is minimized by replacing the machine after 7 years.

B1. The incremental cash flows associated with a 7-year replacement cycle are:

Time	Item	CFBT	CFAT	PV @ 12%
0	I_0	-49,000	-49,000	-49,000
1-7	ΔE	-25,000	-15,000	-68,456
1-5	Depreciation	0	3600	12,977
5	Maintenance	-1000	-600	-340
6	Maintenance	-18,000	-10,800	-5472
7	Salvage (B=4000)	0	1600	724
			TC =	-109,567

$$EAC = 109{,}567\times\left[\frac{0.12\times(1.12)^7}{(1.12)^7 - 1}\right] = \$24{,}008$$

B2. The NPV of \$7.193 million consists of two components: (1) the \$3.999 million NPV for 3 years of production of Clean-e-z and (2) the \$3.194 million NPV of the 6-year Q-10 project with only 1 year of monopolistic pricing power. Because there is only 1 year of monopolistic pricing power in this calculation, only the first year's (ΔR - ΔE) is 5.0 million. The present value is then:

Time	Item	CFBT	CFAT	PV @ 16%
0	Equipment	-6.0	-6.0	-6.000
0	ΔWorking Capital	-0.5	-0.5	-0.500
1-6	Depreciation	0	0.263	0.967
1	ΔR - ΔE	5.0	3.25	2.802
2-6	ΔR - ΔE	3.0	1.95	5.504
6	Salvage (B=1.5)	0	0.525	0.216
6	ΔWorking Capital	0.5	0.5	0.205
			PV =	\$3.194 million

To compute the NPV of 3 years of production of Clean-e-z, note that the first year, since we are waiting to introduce Q-10, revenues minus expenses for Clean-e-z will be at the current level of \$2.6 million. After introducing Q-10, revenues minus expenses for Clean-e-z will drop to \$1.3 million in the second year and because of heavier competition (due to only 1 year of monopolistic pricing power -- versus 2 years) they will drop further to \$0.5 million in the third year.

Time	Item	CFBT	CFAT	PV @ 16%
1-3	Depreciation	0	0.175	0.393
1	ΔR - ΔE	2.6	1.690	1.457
2	ΔR - ΔE	1.3	0.845	0.628
3	ΔR - ΔE	0.5	0.325	0.208
3	Salvage (B=1.5)	0.25	0.688	0.441
3	ΔWorking Capital	0.5	0.5	0.320
			PV =	\$3.447 million

Because we are waiting 1 year to begin production of Q-10, time 0 in the first table above corresponds to time 1 in the second table. To add the two present values, we need to find the value of Clean-e-z production at time 1 so we compound the \$3.447 million present value for 1 year at 16% to get \$3.999 million (= 3.447×[1.16]). The total NPV of this alternative is the sum of the two present values, \$7.193 million (= 3.194 + 3.999).

B3. To decide which machine Y. B. Blue should choose, we compute the EAC for each:

Machine A:

Time	Item	CFBT	CFAT	PV @ 13%
0	I_0	-50,000	-50,000	-50,000
1-6	ΔE	-34,000	-22,100	-88,346
1-6	Depreciation	0	2917	11,660
			TC =	-126,686

$$\text{EAC} = 126{,}686\times\left[\frac{0.13\times(1.13)^6}{(1.13)^6-1}\right] = \$31{,}691$$

Machine B:

Time	Item	CFBT	CFAT	PV @ 13%
0	I_0	-70,000	-70,000	-70,000
1-5	ΔE	-26,000	-16,900	-59,441
1-5	Depreciation	0	4900	17,234
			TC =	-112,207

$$\text{EAC} = 112{,}207\times\left[\frac{0.13\times(1.13)^5}{(1.13)^5-1}\right] = \$31{,}902$$

Y. B. Blue Corporation should choose machine A because it has the lower EAC.

B4. The NPV of \$8.304 million consists of two components: (1) the \$4.144 million NPV for 3 years of production of Clean-e-z and (2) the \$4.160 million NPV of the 6-year Q-10 project with 2 years of monopolistic pricing power.

The analysis for the Q-10 project is identical to the analysis in the text with 2 years of monopoly pricing power. The present value of the 6-year Q-10 project is \$4.160 million.

To compute the NPV of 3 years of production of Clean-e-z, note that the first year we are waiting to introduce Q-10 so revenues minus expenses for Clean-e-z will be at the current level of \$2.6 million. After introducing Q-10, revenues minus expenses for Clean-e-z will drop to \$1.3 million in the second year and \$0.8 million in the third year.

Time	Item	CFBT	CFAT	PV @ 16%
1-3	Depreciation	0	0.175	0.393
1	$\Delta R - \Delta E$	2.6	1.69	1.457
2	$\Delta R - \Delta E$	1.3	0.845	0.628
3	$\Delta R - \Delta E$	0.8	0.52	0.333
3	Salvage (B=1.5)	0.25	0.688	0.441
3	ΔWorking Capital	0.5	0.5	0.320
				PV = \$3.572 million

To add the two present values, we need to find the value of Clean-e-z production at time 1 so we compound the \$3.572 million present value for 1 year at 16% to get \$4.144 million (= 3.572×[1.16]). The total NPV of this alternative is the sum of the two present values, \$8.304 million (= 4.160 + 4.144).

B5. The present value of the costs of the German truck:

Time	Item	CFBT	CFAT	PV @ 12%
0	I_0	-75,000	-75,000	-75,000
1-3	$-\Delta E$	-250,000	-150,000/yr	-360,275
1-3	Depreciation		10,000/yr	24,018
3	Salvage	15,000	9,000	6406
			TC =	-\$404,851

The present value of the costs of the Japanese truck:

Time	Item	CFBT	CFAT	PV @ 12%
0	I_0	-100,000	-100,000	-100,000
1-4	$-\Delta E$	-240,000	-144,000/yr	-437,378
1-4	Depreciation		10,000/yr	30,373
4	Salvage	12,000	7200	4576
			TC =	-\$502,429

$$\text{GERMAN TRUCK EAC} = -404{,}851 \times \left[\frac{0.12 \times (1.12)^3}{(1.12)^3 - 1}\right] = -\$168{,}559$$

$$\text{JAPANESE TRUCK EAC} = -502{,}429 \times \left[\frac{0.12 \times (1.12)^4}{(1.12)^4 - 1}\right] = -\$165{,}417$$

Because it has the lower EAC, Fed Ex should choose the Japanese truck.

B6. a. The present value of the costs of the German truck given the 1-year extension:

Time	Item	CFBT	CFAT	PV @ 12%
0	I_0	-75,000	-75,000	-75,000
1-3	-ΔE	-250,000	-150,000/yr	-360,275
1-3	Depreciation		10,000/yr	24,018
4	-ΔE	-265,000	-159,000	-101,047
4	Salvage	12,000	7200	4576
			TC =	-$507,728

The EAC for the 1-year German truck extension is -$167,162.

The present value of the costs of the German truck given the 2-year extension:

Time	Item	CFBT	CFAT	PV @ 12%
0	I_0	-75,000	-75,000	-75,000
1-3	-ΔE	-250,000	-150,000/yr	-360,275
1-3	Depreciation		10,000/yr	24,018
4	-ΔE	-280,000	-168,000	-106,767
5	-ΔE	-270,000	-162,000	-91,923
5	Salvage	10,000	6000	3405
			TC =	-$606,542

The EAC for the 2-year German truck extension is -$168,261

The present value of the costs of the Japanese truck given the 1-year extension:

Time	Item	CFBT	CFAT	PV @ 12%
0	I_0	-100,000	-100,000	-100,000
1-4	-ΔE	-240,000	-144,000/yr	-437,378
1-4	Depreciation		10,000/yr	30,373
5	-ΔE	-255,000	-153,000	-86,816
5	Salvage	10,000	6000	3405
			TC =	-$590,416

The EAC for the 1-year Japanese truck extension is -$163,787.

The present value of the costs of the Japanese truck given the 2-year extension:

Time	Item	CFBT	CFAT	PV @ 12%
0	I_0	-100,000	-100,000	-100,000
1-4	-ΔE	-240,000	-144,000/yr	-437,378
1-4	Depreciation		10,000/yr	30,373
5	-ΔE	-270,000	-162,000	-91,923
6	-ΔE	-260,000	-156,000	-79,034
6	Salvage	8,000	4800	2432
			TC =	-$675,530

The EAC for the 2-year Japanese truck extension is -$164,306.

The optimal replacement cycle is the one which minimizes the EAC. The 1-year extension is optimal for both the German and Japanese trucks.

b. Fed Ex should purchase the Japanese truck and refit it for 1-year extended life.

B7. a. $r_j = r_f + \beta_j \times (r_m - r_f) = 10\% = 7\% + \beta_j \times (9\% - 7\%)$, so that $\beta_j = 1.5$

b. Because the cash flows are stated in nominal terms, we need to compute the nominal required return before we can compute the NPV:

$$r_n = r_r + i + i \times r_r = 0.10 + 0.03 + 0.03 \times (0.10) = 13.3\%$$

$$\text{NPV} = -40{,}000 + \sum_{t=1}^{5} \frac{\text{CFAT}_t}{(1.133)^t} = \$15{,}427.58$$

c.

$$\text{EAA} = 15{,}427.58 \times \left[\frac{0.133 \times (1.133)^5}{(1.133)^5 - 1}\right] = \$4418.44$$

B8. The **break-even point** is where the total contribution margin exactly equals the total fixed cost of producing a product or service. It is the point at which accounting income is zero. Because the break-even point is based on accounting income it does not consider the time value of money and is not the point at which the NPV of the project is zero. In contrast, the NPV is the present value of the incremental cash flows associated with the project. Because NPV is based on cash flows and applies a discount rate that reflects the nondiversifiable risk of the project, it is the correct measure of project value. If it is not possible to determine a market-based estimate of the required return for an investment, the break-even point might provide an indication of the minimum sales level necessary for the project to undergo further consideration (but remember that the sales level required for a zero NPV virtually always exceeds the break-even sales level).

B9. a. The NPV of the project in real terms:

Time	Item	CFBT	CFAT(real)	PV @ 8%
0	I_0	-250,000	-250,000	-250,000
1-8	ΔR - ΔE	100,000/yr	60,000/yr	344,798
1	Depreciation	0	11,429	10,582
2	Depreciation	0	10,884	9,332
3	Depreciation	0	10,366	8,229
4	Depreciation	0	9,872	7,257
5	Depreciation	0	9,402	6,399
6	Depreciation	0	8,955	5,643
7	Depreciation	0	8,528	4,976
8	Depreciation	0	8,122	4,388
8	Salvage	10,000	10,000	5,403
				NPV = $157,006

b. The nominal cost of capital is 0.08 + (0.08)×(0.05) = 13.4%
Then the NPV of the project in nominal terms is:

Time	Item	CFBT	CFAT(nominal)	PV @ 13.4%
0	I_0	-250,000	-250,000	-250,000
1-8	Depreciation	0	12,000/yr	56,805
1	ΔR - ΔE	105,000	63,000	55,556
2	ΔR - ΔE	110,250	66,150	51,440
3	ΔR - ΔE	115,763	69,458	47,630
4	ΔR - ΔE	121,551	72,930	44,102
5	ΔR - ΔE	127,628	76,577	40,835
6	ΔR - ΔE	134,010	80,406	37,810
7	ΔR - ΔE	140,710	84,426	35,009
8	ΔR - ΔE	147,746	88,647	32,416
8	Salvage	10,000	14,775	5,403
				NPV = $157,006

B10. At 10% per year expected inflation the nominal cost of capital is 0.08 + (0.08)×(0.10) = 18.8%. The change only effects the present value of the depreciation CFATs. It reduces that present value by $9066 to $47,742. So the NPV decreases by the same $9066 to $146,194.

B11. A decision tree differs from the two-state option pricing model developed in Chapter 8 because, in addition to "forks" representing the set of probabilistic outcomes, a decision tree also has "forks" representing managerial decisions.

B12. The potential benefit associated with the use of sensitivity analysis is that it provides an indication of the project's operating leverage and the variation in project NPV that results from varying other parameters. A potential pitfall is that sensitivity analysis may incorporate the impact of diversifiable as well as non-diversifiable risk. It is also important to account for correlations among parameters when combining optimistic and pessimistic estimates for multiple parameters.

B13. First, compute the incremental cash flows:

Time	Item	CFBT	CFAT	PV @ 16%
0	I_0	-16M	-16M	-16M
0	ΔWorking Capital	-3.2M	-3.2M	-3.2M
1-8	ΔR - ΔE	5M	3.25M/yr	14.116M
1-8	Depreciation		0.7M/yr	3.041M
8	Salvage	2M	1.3M	0.397M
8	ΔWorking Capital	3.2M	3.2M	0.976M
			NPV =	-$0.67M

B14. Because of the complexity of the models employed in Monte Carlo simulations and the interrelations among the parameters estimated, it may be virtually impossible to identify the sources of diversifiable risk and eliminate them. As a result, the simulation outcomes are generally based on total risk rather than nondiversifiable risk and may lead to decisions that deviate from the best interests of the shareholders.

B15. a. No. As long as the tax rate is less than 100%, the taxes paid on the revenue will exceed the taxes "saved" on expenses.

b.

Time	Item	CFBT	CFAT	PV @ 16%
0	I_0	-16M	-16M	-16M
0	ΔWorking Capital	-3.2M	-3.2M	-3.2M
1-8	ΔR - ΔE	5M	5M/yr	21.718M
8	Salvage	2M	2M	0.610M
8	ΔWorking Capital	3.2M	3.2M	0.976M
			NPV =	$4.104M

B16. A corporation such as IBM considers many investment projects that make its existing products obsolete. As a result, these projects cannot be analyzed as if they are economically independent from the firm's existing operations. Determining the benefit of introducing a new product must include an analysis of its impact on current products because treating such projects as economically independent would lead to significantly different investment choices.

B17. If the product development has lead to a major advance, the optimal choice is to produce without test marketing. We can see this by working backward on the "major advance" portion of the decision tree. The Figure on the next page illustrates the scenario.

First consider the production decisions in the upper portion of the diagram. The EV of producing is computed for each test marketing result, and decision branches are "pruned": The EV of producing after good results is $3.08M (= 0.6×[5.0M] + 0.3×[1.6M] + 0.1×[-4.0M]). Because this exceeds the zero value of abandoning the project following good test results, we can prune the top abandon branch. Next, the EV of producing after fair test results is $1.50M, which exceeds the zero value of abandoning the project so we prune the second abandon branch. Finally, the EV of producing after poor results is -0.75M, which is less than the zero value of abandoning the project so we prune the produce branch.

The EV of doing the test marketing is the expected value from the above production decisions minus the cost of test marketing the product. EV = 0.5×(3.08M) + 0.3×(1.50M) + 0.2×(0) - 0.25 = $1.74M.

To compute the EV of producing without test marketing, we must first compute the probabilities of high, medium, or low sales. This is done by combining the probabilities from the test marketing branch. For example, the probability of high sales following a good test result is 0.5×(0.6) = 0.3; the probability of high sales following a fair test result is 0.3×(0.3) = 0.09; and the probability of high sales following a poor test result is 0.2×(0.05) = 01. Therefore, the total probability of high sales is 0.4×(= 0.3 + 0.09 + 0.01). Similarly, the total probability of medium sales is 0.4 and the total probability of low sales is 0.2 The EV of producing without test marketing is $2.024M (= 0.4×[5.5] + 0.4×[1.76] + 0.2×[-4.4]), which is greater than the zero value of abandoning the project, so the abandon branch is pruned and the EV of not testing is $2.024M.

Because the EV of producing without testing exceeds the EV of doing the test marketing, the firm should go directly into production if product development has lead to a major advance.

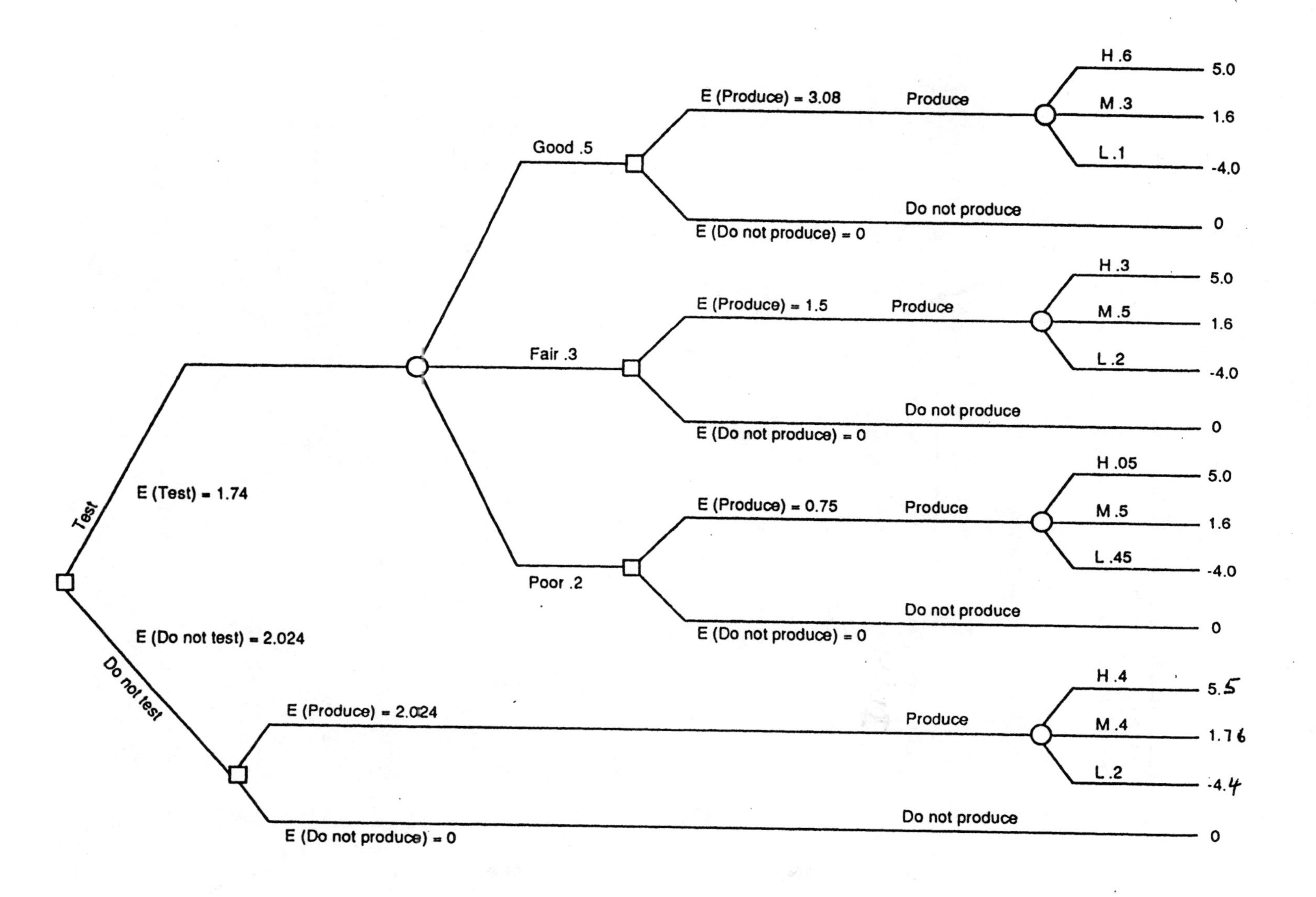
Test
E (Test) = 1.74
E (Do not test) = 2.024
Do not test
Good .5
Fair .3
Poor .2
E (Produce) = 3.08
Produce
E (Do not produce) = 0
Do not produce
H .6
M .3
L .1
5.0
1.6
-4.0
0
E (Produce) = 1.5
Produce
E (Do not produce) = 0
Do not produce
H .3
M .5
L .2
5.0
1.6
-4.0
0
E (Produce) = 0.75
Produce
E (Do not produce) = 0
Do not produce
H .05
M .5
L .45
5.0
1.6
-4.0
0
E (Produce) = 2.024
Produce
E (Do not produce) = 0
Do not produce
H .4
M .4
L .2
5.5
1.76
-4.4
0

B18. a.

Time	Item	CFBT	CFAT	PV @ 15%
0	Salvage	2,500,000	2,500,000	2,500,000
1-10	ΔR - ΔE	?	?	?
1-10	Depreciation (lost)	0	-40,000/yr	-200,751
			NPV =	0

To have a break-even CFAT, the project must produce a present value of $2,299,249 (= 2,500,000 - 200,751). The requires an annual CFAT of $458,130. [PV = 2,299,249, n = 10, r = 15, FV = 0, solve for PMT = 458,130.]

b. A CFAT of $458,130 implies a CFBT of $763,550 (= 458,130/[1 - 0.40]). Assuming zero fixed costs and solving Equation (12.4) for Q, yields

Q = CFBT/c = 763,550/25 = 30,542 So, the zero-NPV sales level is 30,542 units.

B19. a. Note that based on the result in Problem (18b) above, unit sales of 25,000 would produce a negative NPV, while unit sales of 50,000 and 75,000 would produce positive NPV's.

For unit sales of 25,000 units:

Time	Item	CFBT	CFAT	PV @ 15%
0	Salvage	2,500,000	2,500,000	-2,500,000
1-10	ΔR - ΔE	625,000	375,000/yr	1,882,038
1-10	Depreciation	0	40,000/yr	200,751
			NPV =	-$417,211

For unit sales of 50,000 units:

Time	Item	CFBT	CFAT	PV @ 15%
0	Salvage	2,500,000	2,500,000	-2,500,000
1-10	ΔR - ΔE	1,250,000	750,000/yr	3,764,076
1-10	Depreciation	0	40,000/yr	200,751
			NPV =	$1,464,827

For unit sales of 75,000 units:

Time	Item	CFBT	CFAT	PV @ 15%
0	Salvage	2,500,000	2,500,000	-2,500,000
1-10	ΔR - ΔE	1,875,000	1,125,000/yr	5,646,115
1-10	Depreciation	0	40,000/yr	200,751
			NPV =	$3,346,866

b. If the variation in NPV's is due to firm-specific factors, these factors are diversifiable and the firm should take the project.

B20. Denis and Denis should be indifferent to selling versus keeping the project at an NPV = 0.

Time	Item	CFBT	CFAT	PV @ 15%
0	Sales Price	?	?	?
1-10	ΔR - ΔE	3,000,000	1,800,000/yr	9,033,784
1-10	Depreciation	0	40,000/yr	200,751
10	Salvage		1,000,000	247,185
			NPV =	0

The sales price which would produce a zero NPV is \$9,481,720 (= 9,033,784 + 200,751 + 247,185).

C1. a.

$$\text{EAA} = 1500\times\left[\frac{0.12\times(1.12)^4}{(1.12^4 - 1}\right] = \$493.85$$

b. NPV = \$4115.43 (= 493.85/0.12)

c. The 4-year rate = $(1.12)^4 - 1$ = 57.352%, and the present value of an infinite annuity due is simply the present value of a perpetuity plus the time zero payment:

PV = 1500 + 1500/0.57352 = \$4115.43

C2. a. The cost will be the \$100,000 outflow at $t = 0$ less the present value of the inflow at $t = 10$:

PV = 100,000 - [100,000/$(1.12)^{10}$] = \$67,803

b. Using your calculator with PMT = 12,000, n = 10, FV = 0, and r = 12, the present value is computed to be 67,803.

c. If the project did not require the net working capital increase, the \$100,000 could have been invested in an alternative investment at the opportunity cost of 12% where it would have earned a stream of \$12,000 payoffs on that investment. As a result, the cost of having the money "tied up" in net working capital (the difference between the \$100,000 and the present value of recovering it after 10 years) must equal the present value of the stream of forgone payoffs on the alternative investment.

C3. The easiest way to approach this problem is to apply the appropriate inflation rates to compute the nominal cash flows associated with the project and discount those at the nominal required return to find the NPV. First, compute the nominal required return:

$$r_n = r_r + i + i \times r_r = 0.08 + 0.05 + 0.05\times(0.08) = 13.4\%$$

Next, the nominal cash flows are computed. Annual depreciation will be \$40,000 (= [200,000-0]/5) and the annual changes in revenues are computed by applying the 3.5% rate of inflation, while the annual changes in expenses are computed by applying the 6% rate of inflation:

ΔR_1 = 100,000
ΔR_2 = 100,000 ×(1.035) = 103,500
ΔR_3 = 100,000 ×$(1.035)^2$ = 107,122.5
ΔR_4 = 100,000 ×$(1.035)^3$ = 110,871.8
ΔR_5 = 100,000 ×$(1.035)^4$ = 114,752.3

ΔE_1 = 25,000
ΔE_2 = 25,000 ×(1.06) = 26,500
ΔE_3 = 25,000 ×$(1.06)^2$ = 28,090
ΔE_4 = 25,000 ×$(1.06)^3$ = 29,775.4
ΔE_5 = 25,000 ×$(1.06)^4$ = 31,561.9

The incremental cash flows associated with the project are summarized in the table below:

Time	Item	CFBT	CFAT (nominal)	PV @ 13.4%
0	I_0	-200,000	-200,000	-200,000.00
0	ΔW	-25,000	- 25,000	-25,000.00
1	ΔR-ΔE	75,000	51,000	44,973.54
2	ΔR-ΔE	77,000	52,360	40,716.79
3	ΔR-ΔE	79,032.5	53,742.1	36,853.22
4	ΔR-ΔE	81,096.4	55,145.6	33,347.14
5	ΔR-ΔE	83,190.4	56,569.5	30,165.95
1-5	Depreciation	0	12,800/yr	44,584.62
5	ΔW	25,000	25,000	13,331.37
			NPV =	$18,972.63

Letter-Fly should undertake the project because the NPV is positive.

C4. Inflation does not create a problem in the evaluation of a capital budgeting project from a theoretical viewpoint. All that is required to incorporate inflation expectations in a capital budgeting analysis is consistency -- either express all cash flows and the required return in nominal terms **OR** express all cash flows and the required return in real terms. Inflation may create tremendous problems in the evaluation of a capital budgeting project from a practical viewpoint. It can be quite difficult to estimate the expected inflation rate to use, particularly when different cash flows experience different inflation rates or inflation affects the required return and the cash flows differently.

C5. Assuming the equipment for both production methods can be replaced indefinitely if the project is successful, Hopkins should proceed with the project using method A. The alternatives and expected values for each production method are shown in the tree diagram on the next page.

To compute the EV of production, we use the following parameter estimates:

Method A
I_0 = 200,000
ΔE = 305,000/yr
ΔD = 40,000/yr for 5 yrs
equipment life = 10 yrs
S_{10} = 0
project salvage = -50,000

Method B
I_0 = 815,000
ΔE = 75,000/yr
ΔD = 163,000/yr for 5 yrs
equipment life = 7 yrs
S_7 = -50,000
project salvage = 75,000

Because the two methods have different economic lives, we must compute the EAA under the two alternative methods to be able to compare them. Then, assuming that the equipment for each production method can be replaced indefinitely if the project is successful, we can compute the NPV of infinite replacement. The incremental cash flows and NPV associated with each production method under each contingency are summarized in the four tables below:

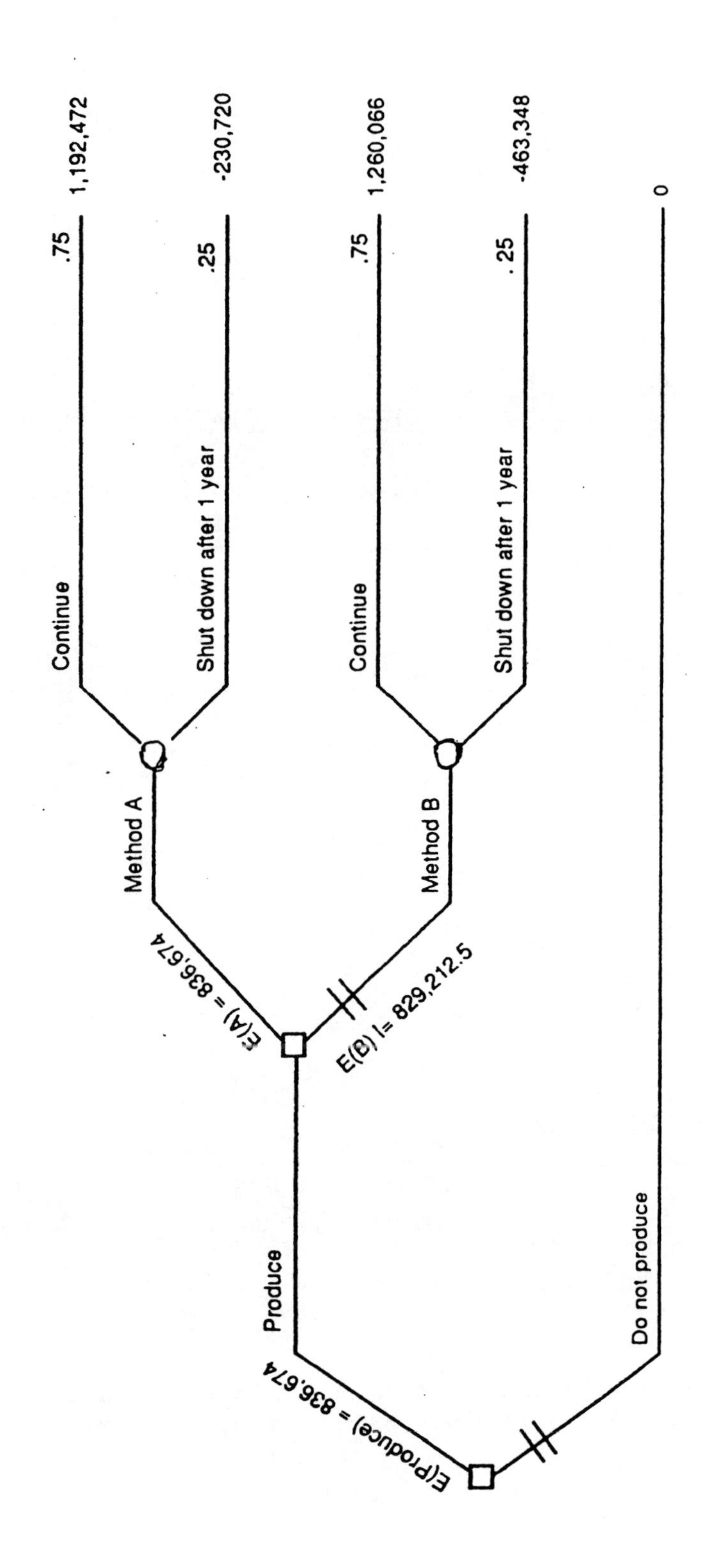
Continue
.75
1,192,472
Shut down after 1 year
.25
-230,720
Continue
.75
1,260,066
Shut down after 1 year
.25
-463,348
0
Method A
Method B
E(A) = 836,674
E(B) = 829,212.5
Produce
Do not produce
E(Produce) = 836,674

Method A and Continuing Operations

Time	Item	CFBT	CFAT	PV @ 18%
0	I_0	-200,000	-200,000	-200,000
1-10	ΔR-ΔE	395,000	248,850	1,118,353
1-5	ΔD	0	14,800	46,282
			NPV =	964,635

$$EAA = 964,635 \times \left[\frac{0.18 \times (1.18)^{10}}{(1.18)^{10} - 1}\right] = \$214,645$$

$$NPV_{\infty} = \frac{214,645}{0.18} = \$1,192,472$$

Method A and Shutdown after 1 year

Time	Item	CFBT	CFAT	PV @ 18%
0	I_0	-200,000	-200,000	-200,000
1	ΔR - ΔE	-125,000	-78,750	-66,737
1	ΔD	0	14,800	12,542
1	Salvage (B=160,000)	-50,000	27,700	23,475
			NPV =	-230,720

Method B and Continuing Operations

Time	Item	CFBT	CFAT	PV @ 18%
0	I_0	-815,000	-815,000	-815,000
1-7	ΔR - ΔE	625,000	393,750	1,500,789
1-5	ΔD	0	60,310	188,600
7	NSV (B=0)	-50,000	-31,500	- 9889
			NPV =	864,500

$$EAA = 864,500 \times \left[\frac{0.18 \times (1.18)^{7}}{(1.18)^{7} - 1}\right] = \$226,812$$

$$NPV_{\infty} = \frac{226,812}{0.18} = \$1,260,066$$

Method B and Shutdown after 1 year

Time	Item	CFBT	CFAT	PV @ 18%
0	I_0	-815,000	-815,000	-815,000
1	ΔR - ΔE	105,000	66,150	56,059
1	ΔD	0	60,310	51,110
1	Salvage (B= 652,000)	75,000	288,490	244,483
			NPV =	-463,348

The EV of each method is the probability of continuing operations multiplied by the NPV of continuing operations plus the probability of shutdown after 1 year multiplied by the NPV of shutdown after 1 year:

$$EV(\text{method A}) = 0.75 \times (1,192,472) + 0.25 \times (-230,720) = \$836,674$$
$$EV(\text{method B}) = 0.75 \times (1,260,066) + 0.25 \times (-463,348) = \$829,212.5$$

So J. Hopkins should proceed with the project using method A.

C6. To find the level of revenues per year that would make Hopkins indifferent to the project using method A, we need to compute the revenue level under continuing operations that makes EV(method A) = 0:

$0.75 \times(\text{NPV of Continuing Operations}) + 0.25 \times(-230{,}720) = 0$

NPV of Continuing Operations = 76,906.67

EAA = 0.18 ×(76,906.67) = 13,843.2

$$\text{NPV} = \frac{13{,}843.2}{\left[\dfrac{0.18\times(1.18)^{10}}{(1.18)^{10}-1}\right]} = \$62{,}213$$

We can put this value in for the NPV at the bottom of the "Method A and Continuing Operations" table and work backwards to find the required revenue level:

Method A and Continuing Operations

Time	Item	CFBT	CFAT	PV @ 18%
0	I_0	-200,000	-200,000	-200,000
1-10	ΔR - ΔE	ΔR - 305,000	0.63×(ΔR - 305,000)	215,931
1-5	ΔD	0	14,800	46,282
			NPV =	62,213

Using this method, the PV of the revenue minus expenses stream must be $215,931. Use your calculator with n = 10, r = 18 and PV = 215,931 to compute the CFAT (PMT) = 48,047.81

CFAT = 0.63 ×(ΔR - 305,000) = 48,047.81; then CFBT = ΔR - 305,000 = 76,266.36; and ΔR = $381,266

C7. The incremental cash flows associated with Ron doing the payroll himself are summarized in the following table:

Time	Item	CFBT	CFAT	PV @ 15%
0	Hardware	-2500	-2500.00	-2500.00
0	Software	-750	-750.00	-750.00
0	Consultant	-4000	-4000.00	-4000.00
0	Setup cost	-2000	-1400.00	-1400.00
1-4	Depreciation	0	543.75	1552.39
1-8	Software update	-75	-52.50	-235.58
1-8	Supplies and time	-480	-336.00	-1507.74
			TC =	- 8840.93

$$\text{EAC} = 8840.93\times\left[\frac{0.15\times(1.15)^{8}}{(1.15)^{8}-1}\right] = \$1970.20$$

Ron should undertake doing the payroll himself because doing so has an EAC of $1970.20 compared with the $2135 (= [1 - 0.3]× 3050) annual after-tax cost of using the CPA firm.

CHAPTER 13 -- CAPITAL BUDGETING IN PRACTICE

A1. Six areas to look for options that might be connected with a firm's capital investment opportunities are (1) the expansion option -- this is the option to expand the production of a current product; (2) the price setting option -- this is the option to change the price of a current product; (3) the abandonment option -- this is the option to terminate a project earlier than was originally planned; (4) the postponement option -- this is the option to postpone an investment project until a later date; (5) the replacement option -- this is the option to replace equipment on alternative replacement cycles; and (6) future investment opportunities -- these arise when undertaking an investment project gives the firm the option to identify additional more valuable investment opportunities in the future.

A2. **Capital rationing** is the process of setting a limit on the amount of new investment that a firm can undertake. The limit can be established directly in dollars or can be achieved indirectly by setting a higher required return.

A3. "Hard" capital rationing occurs when the firm's total new investment cannot exceed the imposed maximum regardless of the existence of additional positive-NPV projects. "Soft" capital rationing occurs when the firm has set a maximum for its total amount of new investment, but depending upon the desirability of the particular set of proposals under review and the firm's condition at the time decisions are actually being made, the firm may invest more than the target amount.

A4. When the technique of zero-one integer programming is applied to the problem of capital rationing, a variable is assigned to each available project. The objective function is to maximize the total NPV of the projects undertaken, subject to the constraint that the total initial investment in projects does not exceed the maximum capital expenditure. The result of the analysis is a variable value of one for projects that should be undertaken and a variable value of zero for projects that should not be undertaken.

A5. By varying the self-imposed capital investment limit, the manager can use capital rationing techniques to analyze the trade-offs among alternative investment sets available to the firm. In this way, the manager can determine the payoffs associated with an increase in the capital expenditure budget.

A6. Reviewing and postauditing the firm's investment decisions is a valuable procedure in that it helps to improve forecasts of expected future cash flows on current and future projects by identifying any consistent biases in past estimates. Another important form of postaudit involves determining the value of abandonment versus continuing the operation of a current project. Some pitfalls associated with review and postaudit are that it may be impossible to evaluate the opportunity costs of forgone alternatives and options after the fact and it may be impossible to identify and measure the incremental cash flows that actually resulted from a decision.

A7. An alternative to keeping a product's price fixed and expanding production to meet the demand at that price is to raise the price of the product to the maximum level that will support the current level of production. As a result, the value of this price setting option should always be considered when evaluating expansion decisions.

A8. Consideration of abandonment value may be more important for a firm engaging in capital rationing than it is for firms that simply take on all positive-NPV projects because firms engaging in capital rationing incur an opportunity cost whenever they pass up positive-NPV projects. Some of these forgone projects may be more valuable than the firm's current investments and abandoning some current projects may provide funds (sale or salvage proceeds) that can be used to expand the firm's capital expenditure budget.

A9. To find the price at which RMC is indifferent between the expansion and an increase in the price, we need to find the price for which the present value of the increased annual net CFAT is equal to the NPV of the expansion alternative. Setting this present value equal to \$342,266, the annual payment for 10 years at a 16% required return must be \$70,815. This corresponds to a before-tax revenue increase of \$118,025 (= 70,815/0.6) or \$0.0118025 per bar. Therefore, RMC is indifferent between the expansion and an increase in price to approximately \$0.624 per bar.

A10. Six factors that can be especially difficult to quantify for inclusion in a capital budgeting NPV calculation are: (1) options -- benefits and opportunity costs associated with options are difficult to identify and quantify (the discussion in problem A1 gives some areas to look for these options); (2) economic dependence -- when the decision to undertake an investment project affects other projects or existing operations, those effects must be included in the analysis of the project; (3) assigning probabilities to possible outcomes -- when decisions depend on the likelihood of future events, it is important to have accurate assessments of those probabilities even though identifying and forecasting the probability of future outcomes is extremely difficult; (4) existing working relationships with suppliers -- such relationships can help to avoid time delays in a particular project; (5) particular expertise -- when current employees have expertise that applies to a project, the value of the project may be increased (likewise, if current employees lack the expertise necessary for the project, its cost may be increased); and (6) experience with quality -- knowledge about the quality of materials from particular manufacturers may add to the value of an alternative involving those materials.

B1. Because of the difference in the risk of the two processes, we must compare the NPVs (including the revenues) rather than the total costs of the alternatives to make this decision. The incremental cash flows for each process are summarized in the tables below:

Process A:

Time	Item	CFBT	CFAT	PV @ 15%
0	I_0	-1,560,000	-1,560,000	-1,560,000
1-8	Depreciation	0	58,500	262,508
1-10	$\Delta R - \Delta E$	560,000	392,000	1,967,357
			NPV =	\$669,865

Process B:

Time	Item	CFBT	CFAT	PV @ 12%
0	I_0	-390,000	-390,000	-390,000
1-8	Depreciation	0	14,625	72,652
1-10	$\Delta R - \Delta E$	365,000	255,500	1,443,632
			NPV =	\$1,126,284

Adam should undertake the expansion using process B because it has the higher NPV.

B2. The IRR is the rate of return that solves the following equation:

$$0 = -100 + \frac{250}{(1 + IRR)} - \frac{156}{(1 + IRR)^2}$$

By trial and error, we find that IRR equals 20% and 30%. A calculator may return either of these answers or an error message.

B3. Tell the firm to reexamine its analysis very carefully. By the Principle of Two-Sided Transactions, the firm that owns the assets will not be willing to sell them for less than they are worth. If the large positive NPV is driven by the ability to purchase the assets at a "bargain," the firm analyzing the project may have overestimated the value of the assets and, thus, the NPV of the project. Alternatively, if the purchasing firm has indeed made a new discovery, the selling firm may raise the price on any subsequent asset purchases, thereby expropriating some of the purchasing firm's NPV. A long-term supply contract might be used to avoid, or minimize the probability of, subsequent price increases.

B4. Because the machines have the same risk and lead to the same increase in revenues, comparing total costs is appropriate. Because they have different economic lives, the EAC should be considered:

Machine A:

Time	Item	CFBT	CFAT	PV @ 20%
0	I_0	1500	1500.00	1500.00
1-4	-Depreciation	0	-43.75	-113.26
1-4	ΔE	475	308.75	799.27
4	-Salvage (B=1000)	-1000	-1000.00	-482.25
			TC =	$1703.76

$$EAC = 1703.76 \times \left[\frac{0.20\times(1.20)^4}{(1.20)^4-1}\right] = \$658.14$$

Machine B:

Time	Item	CFBT	CFAT	PV @ 20%
0	I_0	1500	1500.00	1500.00
1-5	-Depreciation	0	-70.00	-209.34
1-5	ΔE	460	299.00	894.19
5	-Salvage (B=500)	-500	-500.00	-200.94
			TC =	$1983.91

$$EAC = 1983.91 \times \left[\frac{0.20\times(1.20)^5}{(1.20)^5-1}\right] = \$663.38$$

B5. "Hard" capital rationing would be appropriate, for at least a limited time, when market imperfections such as asymmetric information and transaction costs make it costly to raise the necessary funds to undertake all available positive-NPV projects or when the firm is temporarily unable to hire qualified managers for the proposed project.

B6. Trying various combinations of projects or using a zero-one integer programming technique allows us to identify the optimal set of projects: 5, 7, 9, 10, 11, 12, 14, 15. The cumulative outlay for this set is $782,000 providing a total NPV of $878,500.

B7. Both statements are correct. The value of a project is its NPV, which, when measured correctly, includes the present value of the options associated with that project; however, the usual computational methods applied to compute a project's NPV do not include those options. The second formulation reminds us that the true NPV of the project is the NPV as we usually calculate it *plus* an adjustment for the value and cost of any options associated with the project.

B8. Again, trying various combinations of projects or using a zero-one integer programming technique allows us to identify the optimal set of projects: 1, 2, 5, 7, 8, 9, 10, 11, 12, 13, 14, 15. The cumulative outlay for this set is $1,763,000 providing a total NPV of $1,493,500.

B9. a.

Project	Initial Cost	NPV	Profitability Index
A	10,000	1100	0.110
B	20,000	3600	0.180
C	20,000	800	0.040
D	15,000	1600	0.107
E	30,000	4000	0.133
F	40,000	3000	0.075
G	20,000	1400	0.070

b. Choose projects A, B, D, E, and F, for a total initial outlay of $115,000 and total NPV of $13,300.

c. Choose projects A, B, D, E, and G, for a total initial outlay of $95,000 and total NPV of $11,700.

B10. One possible explanation is that upper management has a better understanding of the economic dependencies among your project, other available projects, and existing operations and, upon considering those dependencies, have determined that the project's true NPV is negative. In this case, they are acting in the best interest of the firm's shareholders by not funding the project. Another possibility (not in the best interest of the shareholders) is that upper management may be choosing investment projects "conservatively" to reduce the personal risk of their undiversified human capital.

B11. The NPV of abandonment is $33,333.

Time	Item	CFBT	CFAT	PV @ 15%
0	Bldg & land sale	700,000	640,000	640,000
0	Equipment sale	55,000	60,000	60,000
1-∞	EAC-Bldg (saved)	40,000	30,000	200,000
1-∞	ΔR - ΔE	-200,000	-130,000	-866,667
			NPV =	$33,333

B12. Estimating the value of future investment opportunities for inclusion in capital budgeting NPV calculations requires a number of steps, all of which are extremely difficult: the nature of the opportunities must be identified, the timing and magnitude of their value if they do occur must be determined, and the probability of that occurrence must be accurately estimated. Given the difficulty of estimating the value of future investment opportunities, while it is important to consider the value of such opportunities when valuing investment projects, it is also important not to rely too heavily on this added value to "make" a potential investment attractive.

B13. The incremental cash flows associated with the new investment are summarized in the table below:

Time	Item	CFBT	CFAT	PV @ 17.31%
0	$-I_0$	-100,000	-100,000	-100,000
0	$-\Delta W$	-25,000	-25,000	-25,000
1-8	Depreciation	0	3500	14,582
1-2	$\Delta R - \Delta E$	35,000	22,750	35,925
3-8	$\Delta R - \Delta E$	52,500	34,125	88,287
8	Salvage (B=20,000)	65,000	49,250	13,732
8	$+\Delta W$	25,000	25,000	6970
			NPV =	$34,496

B14. a. No. $5 million per year forever at a cost of capital of 15% is worth $33.33 million at time *t* = 3. There is a 0.40 probability of Upjohn getting the $33.33 million and a 0.60 probability of Upjohn getting nothing, so the expected value is $13.33 million (= (0.60)(0) + (0.40)(33)) at t = 3. The present value of this amount is $8.77 million (= $13.33/(1.15)^3$). Therefore, even with the added value from the growth option the NPV is still negative because the added value is less than -$10 million in DCF-NPV.

b. The minimum option value is $10 million to make the NPV zero.

c. $5.703 million per year. To have a present value of $10 million, Upjohn would need an expected value of $15,208,750 (= $(1.15)^3$10 million). With 0.40 probability of success, the perpetuity would have to be worth $38,021,875 (= 15,208,750/0.40). At a 15% cost of capital, the perpetuity would therefore have to be $5.703 million per year (= (0.15)38,021,875).

B15. a. Because the machines have different economic lives, but will be replaced regularly, Chrysler should buy the machine with the NPV that provides the highest EAA, which is machine A:

Machine A:

Time	Item	CFBT	CFAT	PV @ 12%
0	I_0	-100,000	-100,000	-100,000
1-5	ΔR	85,000	55,250	199,164
1-5	Depreciation	0	7,000	25,233
1-5	$-\Delta E$	-40,000	-26,000	-93,724
			NPV =	$30,673

$$\text{EAA} = 30{,}673 \times \left[\frac{0.12\times(1.12)^5}{(1.12)^5 - 1}\right] = \$8509$$

Machine B:

Time	Item	CFBT	CFAT	PV @ 12%
0	I_0	-36,000	-36,000	-36,000
1-6	ΔR	85,000	55,250	227,155
1-6	Depreciation	0	2,100	8,634
1-6	$-\Delta E$	-62,000	-40,300	-165,690
			NPV =	$34,099

$$\text{EAA} = 34{,}099 \times \left[\frac{0.12\times(1.12)^6}{(1.12)^6 - 1}\right] = \$8294$$

b. The NPV of machine A at a 15% cost of capital is $21,516, and provides an EAA of only $6418. Therefore, under these conditions, Chrysler should buy machine B.

B16. One possibility is that the research team needs the project to maintain the size of its operations to avoid losing staff. To make sure the project is undertaken, the team could inflate the value of the growth option. Steps Upjohn could take to reduce the chance of such self-interested behavior include (i) careful scrutiny and analysis of such "add-ons" to capital budgeting projects, and (ii) a consistent use of postaudits wherein forecasters are held accountable for their forecasts.

C1. As a general rule, it is not likely that a firm can "limit its losses" by selling off a subsidiary that is doing very poorly because of the Principle of Two-Sided Transactions. The potential buyer will be willing to pay only what the subsidiary is worth to them *now*. This is generally not any more than what the subsidiary is now worth to its current owner. Therefore, in most cases, the loss has already been incurred and will not be "limited" by selling the subsidiary. Selling off a subsidiary may be attractive if another firm can bring improved expertise to the subsidiary, thereby increasing the subsidiary's value.

C2. As we discussed in Chapter 9, managers' goals of salary, power, and status are all positively correlated with the size of the firm. A larger firm size is associated with a decreased chance of takeover (which increases management job security) and an increased ability to generate funds internally (which decreases management reliance on the marketplace). For these reasons, it may be in the managers' best interest (and not necessarily in the shareholders' best interest) to forgo the price setting option and choose an expansion project.

C3. Four broad factors that can cause the NPV of a capital budgeting project to be incorrectly measured are: (1) difficulty in defining the incremental cash flows--this is especially relevant in cases of economic dependence; (2) difficulty in determining the correct required return-- it is crucial that the required return reflect the risk of the proposed investment and not necessarily the average risk of the firm; (3) difficulty in predicting the likelihood of future events, such as technological advances that would make the investment obsolete; and (4) difficulty in determining the value of options associated with the investment, such as price setting, future investment, abandonment, and postponement options.

C4. As illustrated in Figure 8-12, additional options do not always add their stand-alone value to the project because of interaction with other options. Many capital budgeting options are not costly to the firm but, rather, are simply inherent in the situation, as is the case with many "hidden" options. For example, the price setting and postponement options often interact with each other and the expansion option, such as when the exercise of one option eliminates one or more other options. When one option "covers" the firm for a number of outcomes (for example, lower or higher than expected sales), additional options may not add additional value. The essential point here parallels our point on Section 13-8 about "nebulous add-ons" to a project: a long list does not substitute for a large value. Therefore, when assessing the value of capital budgeting options, be careful that you don't add a particular value more than once.

C5. Trying various combinations of projects or using a zero-one integer programming technique allows us to identify the optimal set of projects: 1, 5, 7, 8, 9, 10, 12, 15. The cumulative outlay for this set is $1,195,000 providing a total NPV of $1,158,000.

C6. The existence of market imperfections such as tax asymmetries, information asymmetries, and transaction costs may impose costs that were not considered in the original investment project evaluations. In this case, capital rationing may not be an arbitrary imposition of spending limits but, instead, an adjustment of the project NPVs to reflect these additional costs. For example, informational asymmetries in the capital market may make it more costly to raise external equity for an investment project and informational asymmetries in the labor market may make it difficult to hire qualified project managers. Transaction costs also decrease the value of a project that requires special financing transactions.

C7. Price = $(2{,}538{,}000/\text{Demand})^{1/3}$. Let P = Price and D = Demand, then

total revenue $= P \times D = (2{,}538{,}000)^{1/3} \times D^{2/3}$

Marginal Revenue (MR) = the first derivative of total revenue with respect to demand:

MR $= d[\text{Total Revenue}]/dD = [2/3]\times(2{,}538{,}000)^{1/3} \times D^{-1/3}$

Therefore, MR = MC when

MR $= [2/3]\times(2{,}538{,}000)^{1/3} \times D^{-1/3} = 0.387$

Solving for D gives $D = \{[2/3]\times(2{,}538{,}000)^{1/3}/0.387\}^3 = 12{,}974{,}330$

Finally, putting demand into our equation for price

Price = $(2{,}538{,}000/12{,}974{,}330)^{1/3} = 0.5805$

CHAPTER 14 -- CAPITAL MARKET EFFICIENCY: EXPLANATION AND IMPLICATIONS

A1. The capital markets were originally created to bring together and to provide liquidity to the users and suppliers of capital.

A2. The zero-coupon bond cited in the text costs $100 today and pays $1000 at its maturity in 20 years. Using a calculator to find the APY, input PMT = 0, PV = 100, FV = 1000, $n = 20$, and then solve for $r = 12.202\%$. The rate we have solved for, 12.202% is the implied APY on the bond. At the end of one year, the accrued interest on the bond will be $100×0.12202 = $12.20.

To determine the bond's remaining interest in its last year before maturity, we can first to determine the bond's value as of the end of year 19. This can be found by inputing PMT = 0, PV = 100, $n = 19$, and $r = 12.202$. Then solve for FV = 891.25 (because of rounding your calculator may display a slightly different answer). The last year's interest on the bond is, therefore, 1000 - 891.25 =$108.75.

An alternative method for finding the last year's implied interest would be to determine the end-of-year 9 present value of a bond paying $1,000 in one year given an APY of 12.202%. Input PMT = 0, PV = 100, $n = 1$, and $r = 12.202\%$. Then solve for FV = 891.25, the same answer found in the prior method.

A3. The existence of public securities prices provides an inexpensive, fast, and accurate method of estimating the price at which an asset can currently be bought or sold. In this way, the public prices provide an estimate of the asset's "fair market value."

A4. **Riskless arbitrage** is buying (selling) an asset in one market and immediately reselling (buying) it at a higher price in another market.

A5. A perfect market is a market in which there are never any arbitrage opportunities.

A6. Zero NPV does *not* mean zero profit; zero NPV means earning a fair return for the effort expended and the risk taken.

A7. Financial assets are said to be very similar because they are alike in most aspects except for their risk and return characteristics (particularly in comparison with physical assets).

A8. **Collective wisdom** means that, as a group, the work and competition of financial players lead to prices that reflect the probabilities of future events more accurately than any single player could.

A9. It is important to distinguish between anticipated and unanticipated new information because the Principle of Capital Market Efficiency implies that prices will reflect the portion of new information that is anticipated even before the information is announced.

A10. The **value-additivity principle** holds that the value of two or more assets combined is equal to the sum of their respective individual values.

A11. When one party records a dollar of revenue and another party records a dollar of expense in a single transaction, **asymmetric taxes** means that these two parties pay taxes at different marginal rates.

A12. **Asymmetric information** is information known to some but not all parties. Its occurrence means that the information relevant to pricing securities is not costless and available to everyone -- some financial players have more or better information than others or can obtain their information at a lower cost.

A13. Three types of capital market imperfections that may affect corporate decision making are (1) asymmetric taxes -- transactions that lead to a lower total tax bill for the two parties involved -- may provide an explanation for transactions that would otherwise appear to be a zero-sum game; (2) asymmetric information -- the ability to use financial transactions to signal information -- may also explain what appear to be zero-sum games; and (3) transaction costs -- even though transaction costs are symmetrical and don't bias prices upward or downward, their existence can cause financial players to prefer one type of transaction to another.

B1. Six "side" benefits to the existence of capital markets are (1) increased liquidity -- capital markets provide an active and readily available forum for the exchange of securities; (2) decreased total transaction costs -- arranging securities transactions without the capital markets would require more time and money; (3) providing a measure of value -- publicly available transaction prices provide a good estimate of each security's value; (4) the ability to "smooth out" an individual's consumption needs over a lifetime of uneven income; (5) the improved access of "small" participants to investments in real assets; and (6) an increase in the efficiency of total societal investment in real assets.

B2. Competition among people engaged in **arbitrage** is an important factor contributing to capital market efficiency. When one investor discovers an arbitrage opportunity and trades securities to exploit that opportunity, other investors will become aware of the opportunity, and competition among them will drive the security prices up or down so as to eliminate the arbitrage opportunity, thereby reinforcing capital market efficiency.

B3. False. Capital market efficiency and competition among arbitrageurs drives the net present value of being an arbitrageur to zero, which means that they earn a fair return for the effort expended and the risk taken.

B4. An increase in the liquidity of a financial security can appear to be a violation of value additivity because the value of the asset is increased. But value additivity applies when *all* parts are considered, including the loss due to "friction costs." The value of the asset is higher and the loss due to friction is lower, so value is conserved.

B5. The similarity of assets contributes to the efficiency of the capital markets by reducing the evaluation of financial securities to considerations involving risk and return. Investors trading to increase the return on their investments (for a given level of risk) will identify the same opportunities that an arbitrageur would, thus increasing the competition for those opportunities and increasing the efficiency of the market.

B6. The problem of **adverse selection** occurs when simply being in a particular market conveys negative information about the participant. Because high-quality participants are driven out of the market, only low-quality participants remain so that being in the market really is a negative signal; thus, adverse selection is a "self-fulfilling phenomenon."

B7. Inductive reasoning requires inferring the cause from the result. When multiple causes lead to the same result, it is extremely difficult to apply the Signaling Principle and determine the *actual* cause.

B8. On the contrary, random price movements are the natural consequence of efficient capital markets. Because prices in an efficient market react quickly and completely to the arrival of new information and because information arrival is random, the movement of prices must also be random.

B9. According to the law of value conservation, value is neither created or destroyed when assets (cash flows) are combined or separated or exchanged.

B10. Consider two securities, A and B. If it were always true that V(A + B) > V(A) + V(B), then an investor could earn riskless profits by purchasing shares of A and shares of B and selling them as packages containing one share of each.

B11. Transactions costs tend to be symmetrical for a particular type of transaction (for example, stock sale or purchase). Therefore, although transactions costs inhibit arbitrage, they don't bias prices to be higher ore lower. However, transactions costs vary substantially across different types of transactions (for example, a phone versus n-person transaction). As a result, existing differences in transactions costs favor some types of transactions over others, as with larger less frequent transactions versus smaller more frequent transactions.

Asymmetric information and taxes, on the other hand, exist within a particular transaction or transaction type. They can encourage or discourage transactions and push prices higher or lower.

B12. The blessing would be that you would never have to worry about whether you were paying a fair price because all prices would be fair. The curse would be that absolutely no positive-NPV investments would ever exist.

B13. We have seen how the Principles of Self-Interested Behavior, Two-Sided Transactions, and Signaling lead to the Principle of Capital Market Efficiency. One implication of an efficient capital market is that considerations about securities are reduced to considerations about risk and return, and the Principle of Risk Aversion means that for a given level of risk, investors always prefer higher return. As a result, all securities of the same risk must provide the same return -- the return of the best alternative investment of equivalent risk. For the firm, the return is the price it pays for the use of the capital, so the best alternative return is its opportunity cost of capital.

B14. Compute the semi-monthly rate using your calculator and n = 120 (= 2×12×5), PMT = 300.99, and PV = 20,000. This yields the rate 1.1% per half month. APR = 24×(1.1) = 26.40%, and APY = $(1 + .01)^{24}$ - 1 = 30.03%.

B15. It is the variable transaction costs that are relevant in determining whether a particular price differential provides an arbitrage opportunity; the fixed cost is relevant in determining whether it is profitable to enter or stay in the arbitrage business. Thus, if the arbitrageurs have relatively large fixed and small variable transaction costs, more price differentials would be identified as arbitrage opportunities and acted upon, promoting capital market efficiency. If, on the other hand, arbitrageurs have relatively small fixed and large variable transaction costs, few price differentials would be large enough to cover the variable transaction costs, and we would expect more significant departures from capital market efficiency.

C1. You make a series of 12 annual \$10,000 payments beginning at $t = 0$ and ending at $t = 11$. You receive a perpetuity of \$1000 per month beginning at the end of the first month following $t = 11$. The future value (at $t = 11$) of the payments you make must equal the present value (at $t = 11$) of the perpetuity you will receive. The present value of the perpetuity is given by:

$$PVA_{\infty} = \frac{1000}{r_m}$$

where $r_m = (1 + APY)^{1/12} - 1$

The value at $t = 11$ of your payments is equal to the future value of an 11-year ordinary annuity plus the future value of the payment you make today:

$$FV_{11} = 10{,}000 \times \left[\frac{(1 + APY)^{11} - 1}{APY}\right] + 10{,}000 \times (1 + APY)^{11}$$

This is the same value as FVA_{12}, the future value of a simple 12-year annuity. This is because the money does not start earning interest until the first payment is made. That is, the payment stream is a 12-year annuity starting at time $t = -1$. You can create a time-line to illustrate this equivalence. As a result, the APY is the rate that solves the following equation:

$$10{,}000 \times \left[\frac{(1 + APY)^{12} - 1}{APY}\right] = \frac{1000}{(1 + APY)^{1/12} - 1}$$

This equation is solved by trial and error or with a calculator to give an APY of 6.94237%:

r	Left-hand side	Right-hand side
10.00%	\$213,842.84	\$125,405.37
5.00	159,171.27	245,451.54
7.50	184,237.28	165,428.19
7.00	178,884.51	176,861.40
6.94237	178,278.93	178,284.99

C2. Computing the present value of the perpetuity gives the value one period before the first payment ($t = -9$ months) so that value must be compounded 0.75 (= 9/12) of a year to find the value at $t = 0$:

$$PV = \frac{10{,}000}{0.20} \times (1.20)^{0.75} = \$57{,}326.57$$

C3. a. By the definition of perfect capital markets, it is true that the existence of transaction costs means that the market is imperfect. In practice, however, because transaction costs are symmetrical, they do not bias security prices upward or downward and they do not provide an incentive for making a transaction. As a result, transactions costs generally do not cause large deviations from the prices that would be observed in a perfect market.

b. The larger are proportionate transaction costs, the larger can be the bid-ask spread. This is because market participants must buy or sell net of transaction costs and their bids and offers will reflect this fact. In terms of market efficiency, larger proportionate transaction costs mean that market prices can vary more widely before profitable arbitrage transactions (that will "correct" the market price) are possible.

C4. The **opportunity cost** of an alternative is the difference between the value of the alternative and the value of the best possible alternative. Essentially, it is the cost of forgoing the best alternative. For assets of a given risk level, the best alternative is the alternative with the highest rate of return and all investors will choose that alternative. As a result, all assets with the same risk must provide the same of return, and that required return is the appropriate discount rate for computing the present value of the asset.

C5. The net present value calculation accounts for the opportunity cost of alternative investments because the discount rate used for computing the present value of the expected future cash flows is the required return for alternative investments of similar risk.

C6. The Principle of Capital Market Efficiency results in many ways from the other principles. The actions of self-interested market participants (purchases and sales of securities) provide information. Other self-interested market participants stand ready to interpret that information and free-ride whenever they can by applying the Behavioral Principle, and imitating the actions of others. The interaction facilitates the flow of information and helps to enforce the Principle of Capital Market Efficiency.

CHAPTER 15 -- WHY CAPITAL STRUCTURE MATTERS

A1. Essentially, the **corporate tax view** of capital structure holds that the firm should use the largest amount of debt financing that is possible because interest payments are tax-deductible while dividends are not. The **personal tax view** notes that interest income is taxed at higher rates than capital gains income. From the securityholder's perspective this tax disadvantage to interest income offsets the tax benefit to the firm of issuing debt. Thus, the combination of personal and corporate taxes results in leverage having less, if any, effect on firm value.

A2. The **personal tax view** of capital structure mitigates the **corporate tax view** because, while debt interest payments enjoy a tax advantage over equity dividend payments at the corporate level, interest income is taxed at a higher effective personal tax rate than equity income for most investors.

A3. A tax-timing option occurs with capital gains taxes because taxes are paid on capital gains only in the year that the owner sells the asset. That tax can thus be postponed at the owner's option by retaining ownership. This option is valuable because the ability to postpone the payment of the tax reduces its present value.

A4. Restrictive debt covenants can reduce the likelihood of asset substitution by allowing the firm to make investments only in projects in specified industries. By reducing the possibility of asset substitution, debt covenants reduce the agency costs between shareholders and bondholders.

A5. New debt issues provide a monitoring function for a firm's shareholders because when a firm issues new debt, prospective debtholders analyze the firm very carefully to determine a fair price to offer for the new debt. This analysis provides the shareholders with a "free outside audit" of the firm.

A6. Two examples of how transaction costs might affect a firm's choice of financing are (1) the lower issuance costs of debt relative to equity might cause a firm to prefer debt if the decision has been made to obtain additional outside financing and (2) the existence of transaction costs and agency costs causes firms to prefer internal equity financing (that is, retained earnings) over external equity financing.

A7. Potential new equity investors in a firm might be leery of buying newly issued stock in the firm because of asymmetric information. Existing shareholders will not want to issue new equity if the stock is currently undervalued because the new owners will share in the gain when the market price rises, but existing shareholders will want to issue new equity if the stock is currently overvalued because they would like to have additional owners to share the loss when the market price declines. As a result, a new equity issue may indicate that the stock is overvalued.

A8. A **financial leverage clientele** is a group of investors that prefers one type of security or capital structure over others for a particular reason. Because marginal tax rates vary across investors, the reason for different financial leverage clienteles may be tax-based. In this case, firms with particular capital structures appeal to investors in particular tax brackets. Thus, in spite of the seeming tax advantage to debt, some firms may more profitably issue equity in order to meet the demands of a particular clientele.

A9. False.

A10. False. In fact, it is just the opposite.

A11. Taxes would not be asymmetric if Equation (15.7) holds because the corporate tax advantage of debt would be exactly canceled by the personal tax advantage of equity so that the after-personal-tax costs of debt and equity financing would be equal.

A12. A firm that manufactures a unique product using specialized employee expertise might tend to finance with less debt than an otherwise identical firm that manufactures a generic product because the higher the degree of specialization, the lower the liquidity and the higher the expected bankruptcy costs associated with the use of debt.

B1. Your expected net earnings are \$1680 (= 0.24×7000) and your interest payment is \$500 (= 0.10×5000). Your expected gross earnings are \$2180 (=1680 + 500) and this represents a 18.17% return (= 2180/12,000) on all-equity financing.

B2. The owner of shares of Firm B can profit from the following arbitrage opportunity. First, the investor sells his shares of B for \$650,000 (= 1% of \$65,000,000). Next, he lends (buys bonds with) ½ of that amount (\$325,000) at 12% and uses the remaining \$325,000 to buy shares of Firm A. He is able to buy 1.08333% (= 325,000/30,000,000) of the shares of Firm A. Before the transaction, he earned 1% of the net income of Firm B or \$100,000. After the transaction, he has the same personal capital structure because he is lending an amount equal to the debt portion of his stock holding, and he now earns 1.08333% of the net income of Firm A plus the interest on his lending. His earnings after the transaction are \$69,333 (= 1.08333% of \$6,400,000) plus \$39,000 (= 12% of \$325,000) for a total of \$108,333. This transaction provides \$8,333 (= 108,333 - 100,000) per year in arbitrage profits.

B3. Financial distress adversely affects the value of the firm through its tax credits even when the firm is able to use completely all tax credits via loss carryforwards because of the time value of money. Suppose a firm has \$10,000 of depreciation, a 40% tax rate, and a required return of 10%. If the firm is able to use the tax credit now, it is worth \$4,000 (= 0.40×[10,000]) to the firm. If the firm is unable to use the tax credits until next year due to financial distress, they add only \$3636 (= 4000/1.1) to the value of the firm.

B4. This statement is not necessarily true. An extremely restrictive set of covenants will do more than simply lower the firm's interest cost; very restrictive covenants may also restrict the firm's ability to undertake profitable investments or otherwise limit the firm's ability to take actions that are in the best interests of its shareholders. In this case, the opportunity cost of the forgone investments may more than offset the interest savings provided by the restrictive covenants.

B5. a. Miles's has the following partial income statement:

Income Before Taxes	\$10,000
- Taxes	3500
Income After Taxes	6500

Because the firm is unleveraged, the \$6500 per year belongs entirely to the shareholders, and the value of the firm, V_U, is the value of the equity, E, which can be calculated as the present value of a perpetuity:

$$V_U = E = 6500/0.16 = \$40,625$$

b. Now the firm is borrowing $20,312.5 (half of the $40,625 unleveraged value) at 10%, so the partial income statement will be:

Income Before Interest & Taxes	$10,000
- Interest	2031
Income Before Taxes	7969
- Taxes	2789
Income After Taxes	5180

In this case, the debtholders will receive a perpetuity of $2031 in interest payments and the shareholders will receive a perpetuity of $5180 in dividend payments. The value of the leveraged firm, V_L, is the value of the equity, E, plus the value of the debt, D:

$$_L = E + D = (5180/0.1889) + 20{,}312.5 = \$47{,}734$$

Therefore, firm value has increased by $7109.

Alternatively, the value of the firm will increase by $\$7109 = T\times D = 0.35\times(20{,}312.5)$, since $V_L = V_U + T\times D$.

B6. The view that a firm's total market value is invariant to its choice of capital structure rests on the premise that investors can achieve any desired debt-equity mix on their own at the same cost as a firm and, so, will not pay a premium for a particular capital structure. This premise requires perfect capital markets. Three types of capital market imperfections that can cause the capital structure of a firm to have an effect on the value of that firm are (1) asymmetric taxes, (2) transaction costs, and (3) asymmetric information.

(1) The existence of a corporate tax advantage for debt due to asymmetric corporate/personal taxes would cause a firm's capital structure to affect the value of the firm. In this case, with all else being equal, the firm can maximize its value by using the maximum possible amount of debt in its capital structure.

(2) The existence of issuance costs that are lower for debt issues than for equity issues is an example of how transaction costs would cause a firm's capital structure to affect the value of the firm. In this case, with all else being equal, the firm can minimize its costs and maximize its value by using debt rather than equity when it has decided that it is going to obtain additional outside financing.

(3) A situation in which the managers know more than the investors about their firm's expected future earnings is an example where asymmetric information would cause a firm's capital structure to have an effect on the value of the firm. The manager of a firm with high expected earnings may be able to distinguish his firm from one with low expected earnings by his willingness to issue debt, so issuing debt would increase the value of the firm in this case.

B7. a. The firm should be valued as a perpetuity.

Thus, the value of the equity, $E = 1000/0.20 = \$5000$

b. The expected cash flows after corporate taxes at 30% will be $1000\times(1 - 0.30) = \$700$.
The expected value of equity after corporate tax: $E = 700/0.20 = \$3500$.

c. The firm will have $1400 in debt at 10% and a total interest payment of $140, so the cash flows to shareholders after interest and taxes will be:

(1000 - 140)×(1 - 0.30) = $602 per year.

The value of the firm, V_L, will be the value of the stock, E, plus the value of the debt, D:

$$V_L = E + D = (602/0.2389) + 1400 = \$3920$$

$$L = D/V_L = 1400/3920 = 35.7\%$$

d. The after-personal-tax expected payoff to debtholders will be $88.20 (= 140×[1 - 0.37]) and the after-personal-tax expected payoff to equityholders will be $541.80 (= 602×[1 - 0.10]). The value of the firm will be

$$V_L = E + D = (541.80/0.2389) + (88.20/0.10) = \$3150$$

e. The after-personal-tax expected payoff to debtholders will be $105 (= 140×[1 - 0.25]) and the after-personal-tax expected payoff to equityholders will still be $541.80. The value of the firm will be

$$V_L = E + D = (541.80/0.2389) + (105/0.10) = \$3318$$

f. With no taxes, the expected payoff to debtholders will be $140 and the expected payoff to equityholders will be $860 (= 1000 - 140). The value of the firm will be

$$V_L = E + D = (860/0.2389) + (140/0.10) = \$5000$$

Note that with no taxes, the value of the leveraged firm is equal to the value of the unleveraged firm in part a.

B8. A firm operating in an otherwise perfect capital market with corporate and personal taxes satisfying Equation (15.7) will never issue risky debt because of the possibility that the firm may lose some, or all, of the value of its interest deductions during periods of financial distress. In this case, the possibility of the loss of any tax credits would make any leverage undesirable for the firm.

B9. The **capital markets imperfections view** of capital structure incorporates the three dominant capital market imperfections identified in the text--agency costs, taxes, and costs of financial distress. In this view, capital structure choice emerges as a dynamic process that involves various trade-offs and results in a general preference order among a firm's financing alternative. So, for example, at low levels of leverage the tax benefits to debt financing outweigh the costs of financial distress. As a firm adds new debt, it must reassess both the value-enhancing aspects of debt and the costs of adding new debt. As the firm's investment opportunity set shifts over time, it needs to reevaluate its capital structure in light of the changing investment policy. Holding investment policy constant, the firm selects that capital structure which maximizes firm value given a trade-off among tax benefits from leverage, costs of financial distress, and changes in agency costs.

C1. a. No, the operating head is not correct.

b. While debtholders are promised a fixed payment that provides their entire return, shareholders are proportionate claimants on the residual value of the firm. Therefore, the cash dividends the shareholders receive are only part of their return. The rest of their return is provided by reinvested earnings. Because of the higher risk connected with equity ownership, the required return on common equity (the firm's cost of common equity) is in fact higher than the interest rate paid to the debtholders.

C2. It is true that the required return on equity and the required return on debt both increase with increasing leverage. However, the required return on equity is higher than the required return on debt for all degrees of leverage, and increasing the leverage means that the proportion of lower-cost debt is increasing. In a perfect capital market environment, the effect of substituting debt for equity just offsets the effect of higher leverage on the separate costs of debt and equity; the net result is that the weighted average of the two remains constant.

C3. A firm's debt being larger than its equity does not necessarily imply that the firm is "in trouble." What the firm "owns" is described on the asset side of its balance sheet and the mix of debt and equity simply describes how those assets are financed. The proportion of debt and equity a firm chooses should represent management's attempt to balance the costs and benefits of each to provide a financing mix that maximizes shareholder wealth.

C4. You should take the "great" investment alternative. In a perfect capital market, the standard deviation of the market required return for a given risk level is zero. Although the capital markets are not exactly perfect, the Principle of Capital Market Efficiency implies that the standard deviation of the market required return for a given risk level is relatively small. In contrast, the Principle of Valuable Ideas implies that there are positive-NPV opportunities that occur. That is, it implies that the markets for real assets are not nearly as efficient as the capital markets. Therefore, the standard deviation of expected future cash flows is comparatively large. Since in either case, your outcome is to be 3 standard deviations in your favor, you should take the alternative with the largest standard deviation.

CHAPTER 16 -- MANAGING CAPITAL STRUCTURE

A1. Pro forma analysis is an important prerequisite to choosing a capital structure because it allows the firm to determine the impact of the proposed capital structure on (1) its ability to utilize fully the implied tax shields, (2) its future external financing requirements, and (3) its dividend policy.

A2. The major reason that subordinated debt is typically rated lower than senior debt is that subordinated debt has a greater exposure to default risk.

A3. Interest coverage ratio = EBIT/Interest Expense = 30 million/10 million = 3.0

A4.

$$\text{Fixed-charge coverage ratio} = \frac{\text{EBIT} + (0.3333)\times\text{Rentals}}{\text{Interest Expense} + (0.3333)\times\text{Rentals}} = \frac{30 + 5}{10 + 5} = 2.33$$

A5.

$$\text{Debt-service coverage ratio} = \frac{\text{EBIT} + (0.3333)\times\text{Rentals}}{\text{Interest Expense} + (0.3333)\times\text{Rentals} + \dfrac{\text{Principal Pmts}}{(1 - T)}}$$

Debt-service coverage ratio = [30 + 5]/[10 + 5 + 6/(1 - 0.4)] = 1.4

A6. Selecting a target senior debt rating is a reasonable approach to choosing a capital structure because the debt rating encompasses three of the five principal considerations that affect the capital structure decision. The debt rating accounts for the firm's ability to service its debt, the degree of protection afforded by the liquidity of the firm's assets, and largely determines the firm's ease of access to the capital markets. A target senior debt rating of single-A is a prudent objective because it is the lowest rating that historically has allowed the issuer to maintain uninterrupted access to the capital markets.

A7. If a firm's capital structure is different from its target capital structure, the firm can bring its capital structure in line with its target by adjusting its future financing mix appropriately. The firm can change its capital structure quickly by making an exchange offer, recapitalization offer, debt or share repurchase, or stock-for-debt swap.

a. If the firm believes it is over leveraged, it would want to decrease its leverage with a stock-for-debt exchange.

b. If the firm believes it is under leveraged, it would want to increase its leverage with a debt-for-stock exchange.

A8. The total present value of the investment is $595,000 (= 270,000 + 325,000) and the debt is $200,000, so the expected leverage ratio (*L*) is 0.336 (= 200,000/595,000).

A9. This approach would be extremely cumbersome; transaction costs would outweigh the benefits.

A10. a. Under perfect capital markets,

$$\text{WACC} = (1 - L)\times r_e + L\times r_d = 0.7\times(18) + 0.3\times(8) = 15\%$$

b. With corporate taxes,

$$\text{WACC} = (1 - L)\times r_e + L\times(1 - T)\times r_d = 0.7 \times(18) + 0.3 \times 0.65)\times(8) = 14.16\%$$

A11. Using Equation (16.9):

$$\text{WACC} = 0.43 - 0.20\times(0.52)\times(0.26)\times\left[\frac{1.43}{1.26}\right] = 39.93\%$$

A12. To find the unleveraged required return, r, we use Equation (16.11) where

$$\text{WACC} = (1 - L)\times r_e + L \times(1 - T)\times r_d = 0.5 \times(16) + 0.5 \times 0.6)\times(10) = 11\%; \text{ and}$$

$$H(L) = [T^*\times L \times r_d]/[1 + r_d] = 0.25 \times 0.5)\times 0.10)]/[1.10] = 0.011363636; \text{ so that}$$

$$r = [\text{WACC} + H(L)]/[1 - H(L)] = 0.11 + 0.011363636)/(1 - 0.011363636) = 0.1228 = 12.28\%$$

A13. The NPV of this investment equates the value of the firm with the investment minus the value of the firm without the investment:

$$\text{NPV} = (170{,}650/0.12165) - (140{,}301/0.10) = -\$215$$

Because the NPV is negative, the firm should not undertake the investment.

A14. a. For Epson,

$$H(L) = 0.20\times(0.6)\times(0.12)/(1.12) = 0.012857 \text{ and}$$

$$\text{WACC} = 0.4\times(0.20) + 0.6\times(0.6)\times(0.12) = 12.32\%;$$

so that from Equation (16.11), $r = 0.1232 + 0.012857)/(1 - 0.012857) = 13.78\%$

b. Averaging the estimates for Epson and H-P is a project unleveraged required return of 13.915%.

c. $\text{WACC} = 0.13915 - 0.15)\times(0.3)\times(0.11)\times(1.13915)/(1.11) = 0.13915 - 0.00508 = 13.4\%$

A15. If a firm is unable to utilize fully the tax benefits that result from the tax-deductibility of its interest payments, then there is less incentive to use debt financing, and the firm will have a lower degree of leverage in its optimal capital structure.

B1. a. The following target ranges are recommended:

Fixed-Charge Coverage Target:	3.40 to 4.00
Cash Flow/Total Debt Target:	55 to 65%
Long-Term Debt/Capitalization Target:	25 to 30%

b. Additional considerations include the firm's ability to utilize fully non-interest tax credits and debt management considerations such as issuance costs.

c. The chief financial officer must determine whether Bixton will be able to utilize its larger-than-average research and development and foreign tax credits at debt levels corresponding to the suggested targets.

B2. Research shows that a firm's debt rating is highly correlated with its size. Because of this relationship, a firm undertaking a comparable-firms approach to choosing its capital structure should limit its set of comparable firms to those firms that are approximately its size.

B3. a. The following targets are suggested:

Long-term Debt/Capitalization Target:	40 to 45%
Cash Flow/Long-Term Debt Target:	30 to 40%
Fixed Charge Coverage Target:	2.50 to 2.75

b. Firms A, C, and F have senior debt ratings consistent with Sanderson's target so they are good comparables. Of these three, firm C has total capitalization closest to that of Sanderson's. Firm E is a particularly poor comparable due to the significant size difference.

c. The relatively high ratio of long-term debt to total capitalization and relatively high fixed charge coverage ratio offset each other to some degree. Nevertheless, Sanderson should plan on reducing its long-term debt ratio to the target range within at most a few years.

B4. Lenders would be willing to lend a larger proportion of the market value of tangible assets than of intangible assets because tangible assets tend to hold their value much better in the event of financial distress. For example, there may be an active secondary market for the tangible assets, in which case the expected loss due to financial distress would be smaller.

B5. If the interest coverage ratio > 1, then adding a fixed amount (1/3 of rentals) to both the numerator and the denominator will cause a larger proportionate increase in the denominator and the resulting fraction (the fixed charge coverage ratio) will be smaller. Comparing the fixed charge coverage ratio with the debt service coverage ratio, the two have the same numerator, but the debt service coverage ratio has a larger denominator, so it must be smaller. All three ratios have the same value when the firm has zero rentals and zero principal repayments during the period.

B6. The firm would maintain a lower degree of leverage than a cross section of single-A-rated firms if it is unable to utilize fully its interest tax shields or if debt management considerations such as issuance expenses cause it to issue less debt.

B7. a. Interest coverage ratio = 2.5 (= 100 *million*/40 *million*).

b. In this case, the interest expense will increase by $10 million with no immediate increase in earnings; consequently, the Interest coverage ratio = 2.0 (= 100 million/50 million).

c. In this case, the interest expense will increase by $10 million and the earnings will increase by $8 million. As a result, Interest coverage ratio = 2.16 (= 108 million/50 million).

B8. We first need a loan amortization schedule to identify the interest payments over the life of the loan:

Year	0	1	2	3	4
Balance, start of year	0	100,000	100,000	80,000	50,000
Interest	0	14,000	14,000	11,200	7000
Principal	0	0	20,000	30,000	50,000
Balance, end of year	100,000	100,000	80,000	50,000	0

The net adjusted present value of the project is the APV, from Equation (16.5) minus the initial cost:

$$\text{APV} = \sum_{t=1}^{n} \frac{\text{CFAT}_t}{(1+r)^t} + \sum_{t=1}^{n} \frac{T^{*} \times \text{INT}_t}{[1+r_d]^t}$$

$$= \sum_{t=1}^{4} \frac{150{,}000}{(1.2)^t} + \frac{100{,}000}{(1.2)^4} + \left[\frac{14{,}000}{1.14} + \frac{14{,}000}{(1.14)^2} + \frac{11{,}200}{(1.14)^3} + \frac{7000}{(1.14)^4}\right] \times [0.20]$$

$$= 388{,}310 + 48{,}225 + 6951 = \$443{,}486$$

$$\text{Net APV} = \text{APV} - \text{Cost} = 443{,}486 - 200{,}000 = \$243{,}486$$

B9. Query should adopt the 15% debt financing alternative because it has an opportunity cost of capital of 16.275% (= (1 - 0.15)×(18) + 0.15 ×(1 - 0.35)×(10)) compared with 16.2755% (= (1 - 0.45)×(23.21) + 0.45 ×(1 - 0.35)×(12)) for the 45% debt alternative. The taxes in this environment are approximately symmetrical since the required returns for the two alternatives are nearly identical. There appears to be a slight tax advantage to equity financing.

B10. a. Applying the CAPM,

$$r_e = r_f + \beta_L \times (r_m - r_f) = 8 + 1.45 \times (15 - 8) = 18.15\%$$

b. The after-tax weighted average cost of capital is

$$\text{WACC} = (1 - 0.27) \times (18.15) + 0.27 \times (1 - 0.35) \times (12) = 15.36\%$$

c. From Equations (16.10) and (16.11),

$$H(L) = [0.21 \times (0.27) \times 0.12]/1.12 = 0.006075; \text{ and}$$

$$r = [0.1536 + 0.006075]/[1 - 0.006075] = 16.07\%$$

d. Using the CAPM again with the unleveraged required return and solving for the unleveraged beta gives

$$16.07 = 8 + \beta_U \times (15 - 8); \text{ so that } \beta_U = 1.15$$

B11. a. We apply the dividend growth model and solve for r:

$$r_e = D_1/P_0 + g = 2.22/20 + 0.05 = 16.1\%$$

b. RTE's after-tax weighted average cost of capital is

$$WACC = (1 - 0.35)\times(16.1) + 0.35\times(1 - 0.3)\times(9) = 12.67\%$$

c. Solving Equation (15.9) for r gives

$$r = WACC/(1 - T\times L) = 12.67/(1 - 0.3\times(0.35)) = 14.16\%$$

d. The unleveraged beta implied by $r = 14.16\%$ is given by

$$14.16 = 6 + \beta_U\times(12 - 6);\text{ so that } \beta_U = 1.36$$

e. A leveraged beta of 2.2 implies a leveraged required return on RTE's common stock of

$$r_e = 6 + 2.2\times(12 - 6) = 19.2\%$$

The market model gives the best estimate of the required return on common stock. So we might conclude that the dividend growth model has underestimated the leveraged required return and, in that case, we would want to adjust all of the estimates in parts a-d upward.

B12. First, we use the equity market value of $56 million (4 million shares at $14 per share) and the total debt market value of $45 million ($30 million of long-term bonds plus $5 million of notes payable and $10 million of other current liabilities) to compute the debt ratio, $L = 45/(56+45) = 44.55\%$. There are several methods we can use to estimate the required return on the common stock, including the market model, the dividend growth model, and the P/E ratio method. Each of these estimates is computed below. (You could also use the borrowing rate method--Emery's rule--see footnote12 on page 519 of the text.)

Market model: $r_e = 6 + 1.1\times(13.75 - 6) = 14.525\%$

Dividend growth model:

(1) Estimate g, where $0.72\times(1 + g)^6 = 1.02$; $g = 6\%$ (note: # years = 5 + 1)

(2) $r_e = 1.02/14 + 0.06 = 13.29\%$

P/E ratio method:

(1) $P/E = 14/2.09 = 6.7$

(2) $r_e = 2.09/14 = 14.93\%$

The estimates are reasonably close, and the market model is the most reliable, so we'll use $r_e = 14.5\%$

The weighted average cost of capital then is

$$WACC = (1 - 0.4455)\times(14.5) + 0.4455\times(1 - 0.3)\times(10.05) = 11.1743\%$$

$$H = T^*\times L\times r_d/(1 + r_d) = 0.18\times 0.4455)\times 0.1005)/(1.1005) = 0.007323$$

Finally, the unleveraged required return is

$$r_e = [WACC + H(L)]/[1 - H(L)] = 0.111743 + 0.007323)/(1 - 0.007323) = 12\%;\text{ and}$$

$$\beta_U = (r_e - r_f)/(r_m - r_f) = (12 - 6)/(13.75 - 6) = 0.77$$

B13. For each firm, the market value of the debt (MV D) is computed as the number of bonds multiplied by the price per bond and the market value of the equity (MV E) is computed as the number of shares multiplied by the price per share. L is estimated as the market value of the debt divided by the total market value of the debt and equity.

The semiannual required return for each bond is computed using the current price, the \$1000 future value, the coupon payments, and 20 periods (= 2×10). The APY of the outstanding bonds is:

$$r_d = (1 + r_{s\text{-}a})^2 - 1$$

Ignoring the tax effects associated with the bonds selling at either a premium or a discount, we can use the APY of the outstanding bonds as an estimate of the APY of new debt.

The required return is computed by applying the CAPM: $r_e = r_f + \beta \times (r_m - r_f)$

H(L) is computed by applying Equation (16.10); the WACC is computed by applying Equation (16.4); and the unleveraged required return, r, is computed by applying Equation (16.11). Averaging r for the three comparable firms gives an estimate of the unleveraged required return for the project.

	MV D	MV E	L	r_d	r_e	H(L)	WACC	r
A	11.77M	25M	0.32	0.1064	0.158	.006155	0.1296	0.1366
B	60.30M	60M	0.50	0.1207	0.166	.010770	0.1222	0.1344
C	27.50M	110M	0.20	0.1073	0.162	.003876	0.1435	0.1479

The average r for the 3 firms is 13.96%. Using 13.96% as the unleveraged required return for the project, L = 0.4, and r_d = 11.25%, we can apply Equation (16.9):

$$\text{WACC} = r - T^* \times L \times r_d \times \left[\frac{1+r}{1+r_d}\right] = 0.1396 - 0.2 \times (0.4) \times (0.1125) \times \left[\frac{1.1396}{1.1125}\right] = 13.0\%$$

B14. a. We must first find the interest payment each period:

Time	0	1	2	3
Balance	0	700,000	600,000	400,000
Interest	0	105,000	90,000	60,000
Principal	0	100,000	200,000	400,000
Balance	700,000	600,000	400,000	0

The net APV is computed by applying Equation (19.11) and deducting the initial cost:

$$\text{APV} = \sum_{t=1}^{n} \frac{\text{CFAT}_t}{(1+r)^t} + \sum_{t=1}^{n} \frac{T^* \times \text{INT}_t}{[1+r_d]^t}$$

$$= \sum_{t=1}^{3} \frac{1{,}000{,}000}{(1.25)^t} + \frac{1{,}000{,}000}{(1.25)^3} + \left[\frac{105{,}000}{1.15} + \frac{90{,}000}{(1.15)^2} + \frac{60{,}000}{(1.15)^3}\right] \times [0.25]$$

$$= 1{,}952{,}000 + 512{,}000 + 49{,}702 = \$2{,}513{,}702$$

$$\text{Net APV} = \text{APV} - \text{Cost} = 2{,}513{,}702 - 1{,}900{,}000 = \$613{,}702$$

b. $L = 700{,}000/2{,}513{,}702 = 0.28$, at the start of the project

c. The WACC is the rate that solves the following equation

$$2{,}513{,}702 = \sum_{t=1}^{3} \frac{1{,}000{,}000}{(1 + r^*)^t} + \frac{1{,}000{,}000}{(1 + r^*)^3}$$

By trial and error or using a calculator, we find that WACC = 23.81%.

d. Solve the WACC equation for r_e:

$$\text{WACC} = (1 - L)\times r_e + L\times(1 - T)\times r_d = 23.81 = (1 - 0.28)\times r_e + 0.28\times(1 - 0.30)\times(15);$$

so that $r_e = 29\%$

B15. It is important to note that the required return is not a historical cost of funds because it must reflect the rate at which investors would provide financing today for the project under consideration -- that is, the required return must be the opportunity cost of capital. If the firm's opportunity cost of capital is greater than its historical cost, then using the historical cost as the required return would cause the firm to overvalue the project and might lead it to accept a negative-NPV investment. If the firm's opportunity cost of capital is less that its historical cost, then using the historical cost would cause the firm to undervalue the project and might lead it to forgo a positive-NPV investment.

B16. **Subordinated debt** is debt that ranks behind senior debt in the event of default. Convertible debt is typically issued in the form of subordinated debt rather than senior debt because both issuers and investors expect the issue to be converted into common stock in a relatively short time period. There is little cost saving to issuing convertible debt on a senior basis. Subordinated debt is advantageous to senior debtholders in that, like equity, it provides another layer of protection in the event of default. Subordinated debt is disadvantageous to senior debtholders because the higher interest rate on subordinated debt increases the firm's fixed charges (and hence financial risk) and also because strict priority is not always preserved in a bankruptcy settlement.

C1. Firms with a high degree of operating leverage have larger amounts of tangible fixed assets and a correspondingly larger amount of tax shields associated with depreciation and investment tax credits. As a result, firms with a high degree of operating leverage are less likely to be able to utilize fully the tax benefits associated with debt financing and so have a lower degree of financial leverage. Also, a higher degree of operating leverage implies a greater exposure to the risk of financial distress for any particular degree of financial leverage, which leads to a tendency for firms with greater operating leverage to maintain less financial leverage in order to control their exposure to the risk of financial distress.

C2. a. If the investor owns 5% of the stock of firm A, and this is half-financed with borrowing, the debt ratio for his holding is $L = 0.5\times(1) + 0.5\times(0.25) = 0.625$. To match this position with an investment in the stock of firm B, let x be the proportion of the investment that is financed with borrowing and find x that solves the following equation:

$L = 0.625 = x\times(1) + (1 - x)\times 0.3333)$; rearranging, $0.625 = x + 0.3333 - 0.3333\times x$; and $0.2917 = 0.6667\times x$, so that $x = 0.4375$. Consequently, an investment in 5% of the common stock of B that is financed with 43.75% debt would produce identical returns.

b. If the investor owns 10% of the stock of firm B, none of which is financed with borrowing, the debt ratio of his holding is the debt ratio of firm B, $L = 0.3333$. To match this position with an investment in the stock of firm A, we need to find x that solves the following equation:

$L = 0.3333 = x\times(1) + (1 - x)\times(0.25)$; rearranging, $0.3333 = x + 0.25 - 0.25\times x$; and $0.0833 = 0.75\times x$, so that, $x = 0.1111$

Consequently, an investment that in 10% of the common stock of A that is financed with 11.11% debt would produce identical returns.

C3. Restrictive covenants that disallow things such as leases and liens on the firm's assets act to protect the interest of the bondholders against expropriation by the shareholders. Without such restrictions, a firm's managers could grant liens to one class of lenders to the detriment of the other unsecured lenders or the managers could enter into leases, which represent a form of secured debt. (Note that because missing lease payments forces the firm into default, rating agencies consider leases in their analysis of the firm's financial position.) As a result, the existence of these restrictive covenants prevents a firm's managers from creating a new senior class of debt that would dilute the claims of the firm's existing debtholders. By protecting the existing debtholders, the covenants should allow the firm to achieve a higher rating on its bonds than would be possible without such restrictive covenants.

C4. Disagree. The existence of a viable junk bond market means that firms do not have to maintain a high credit rating for the sole purpose of ensuring uninterrupted access to the capital markets. But access to the capital markets is only one of the important factors affecting a firm's choice of capital structure. Existence of a viable junk bond market does not mean that firms can "comfortably" maintain higher degrees of leverage than they could prior to the development of this market. The higher return on these bonds reflects the other risks inherent in increasing leverage to the point that bond ratings fall below investment grade.

C5. a. If minority interest consists of outstanding common stock of a subsidiary that the parent firm has no intention of repurchasing or otherwise retiring, then it should be treated as common equity of the consolidated firm.

b. If minority interest consists of redeemable preferred stock of a subsidiary, which the firm is obligated to redeem in equal annual amounts over the next 5 years, then it is a fixed obligation of the subsidiary and should be treated as a debt equivalent of the consolidated firm.

Whether minority interest is really debt or equity depends on the nature of the payment obligations that it imposes on the parent firm.

C6. To analyze this problem, we must isolate the amounts that are actual cash flows for Ida:

Time	CFBT
0	-10,000.00
1-60	+2500.00
60	-199,242.62

The required return that makes the NPV of this investment zero is 1.2313% per month. This corresponds to an APY of 15.82%. Ida should not undertake this investment unless she can invest the proceeds elsewhere at a riskless APY higher than 15.82%.

C7. a. The principal payments consist of equal installments so the balance declines by $75,000 each period. The interest payments each period are summarized below:

Time	Beginning Balance	Interest
1	$750,000	$90,000
2	675,000	81,000
3	600,000	72,000
4	525,000	63,000
5	450,000	54,000
6	375,000	45,000
7	300,000	36,000
8	225,000	27,000
9	150,000	18,000
10	75,000	9000

$$\text{APV} = \sum_{t=1}^{10} \frac{400{,}000}{(1.17)^t} + \frac{200{,}000}{(1.17)^{10}} + \left[\sum_{t=1}^{10} \frac{\text{INT}_t}{(1.12)^t}\right] \times [0.30] = \$2{,}002{,}919$$

The net APV is computed by applying Equation (19.11) and deducting the initial cost:
Net APV = 2,002,919 - 1,000,000 = $1,002,919

b. The actual WACC is the rate that solves the following equation

$$2{,}002{,}919 = \sum_{t=1}^{10} \frac{400{,}000}{(1 + r^*)^t} + \frac{200{,}000}{(1 + r^*)^{10}}$$

By trial and error or using a calculator, we find that the actual WACC = 15.683%.
L = 750,000/2,002,919 = 0.3745. We solve the WACC equation for r_e:

WACC = $(1 - L)\times r_e + L\times(1 - T)\times r_d$ = 15.683 = $(1 - 0.3745)\times r_e + 0.3745\times(1 - 0.35)\times$ (12); and

r_e = 20.40%

Alternatively, using Equation (16.9), the rebalancing approximation for the WACC, we get

WACC = $r - T^*\times L\times r_d[(1 + r)/(1 + r_d)]$ = $17 - 0.3\times(0.3745)\times(12)\times(1.17/1.12)$ = 5.592%

This implies a leveraged required return of r_e = 20.25%

Finally, using Equation (15.9), the (corporate tax view) approximation for the WACC,

WACC = $r\times(1 - T^*\times L)$ = $17\times[1 - 0.3\times 0.3745)]$ = 15.090%,

which implies a leveraged required return of r_e = 19.45%

c. The NPV of the project depends on the value used to discount the cash flows. Using the WACC that solves for the APV makes the computed NPV equal to the APV

$$NPV = -1{,}000{,}000 + \sum_{t=1}^{10} \frac{400{,}000}{(1.15683)^t} + \frac{200{,}000}{(1.15683)^{10}} = \$1{,}002{,}919$$

Alternatively, using the rebalancing approximation of the WACC, we get an NPV of

$$NPV = -1{,}000{,}000 + \sum_{t=1}^{10} \frac{400{,}000}{(1.15592)^t} + \frac{200{,}000}{(1.15592)^{10}} = \$1{,}009{,}987$$

Finally, using the corporate tax view approximation of the WACC, we get an NPV of

$$NPV = -1{,}000{,}000 + \sum_{t=1}^{10} \frac{400{,}000}{(1.15090)^t} + \frac{200{,}000}{(1.15090)^{10}} = \$1{,}049{,}692$$

d. When T^* is positive and financing and investment decisions are interrelated, the required return is unique and depends on the pattern of debt payments. The MM (corporate tax view) will always provide an approximation that is less than or equal to the actual rate. (Recall that it is equal to the unique rate in the case of perpetuities.) The rebalancing approximation can be higher or lower than the actual but, because of this, will generally provide a closer approximation to the actual WACC.

CHAPTER 17 -- WHY DIVIDEND POLICY MATTERS

A1. The **tax differential view** of dividend policy advocates low payout ratios. According to this view, shareholders prefer capital gains over dividends because capital gains have typically been taxed less than dividends because of postponed taxes and lower rates.

A2. According to the **signaling view** of dividend policy, dividend changes provide information to investors regarding changes in management's expectations about the firm's future earnings.

A3. A **dividend clientele** is a group of investors that prefers the firms it invests in to follow a particular dividend policy. The existence of dividend clienteles mitigates the tax differential view of dividend policy because, based on their tax positions, investors will have preference for investing in high- or low-cash-dividend stocks. As long as there is a sufficient supply of investment opportunities for each clientele, investors will not pay a premium for stock of a firm that follows a particular dividend policy.

A4. False. Empirical evidence shows that the stock market reacts positively to announcements of dividend increases and adversely to announcements of dividend decreases.

A5. False. Most firms follow a stable dollar amount dividend policy, normally increasing the dividend amount only when a change in the firm's earnings and cash flow prospects justifies a change.

A6. (1) The existence of brokerage commissions tends to favor dividend payments because these commissions make it costly for investors to create "homemade" dividends by selling shares of stock. (2) The existence of flotation costs favors dividend reinvestment over external equity financing as a means of raising additional equity capital.

A7. If a change in tax law affects the optimal dividend policy for a particular clientele and firms following their new optimal dividend policy are in short supply, it may be possible for firms to earn a positive NPV by altering their dividend policies to meet the new demand.

A8. The phrase "bird-in-the-hand *fallacy*" describes the logical flaw in the traditional view of dividend policy, which holds that investors will prefer dividends to capital gains because dividends are cash in hand, while reinvesting that cash to obtain future capital gains is risky. By the Principle of Risk-Return Trade-Off, the difference between dividends now and dividends in the future is a function of risk and the time value of money. To get the dividend now, we must accept a lower return, and, as long as the future capital gains provide appropriate compensation for the risk and the time value of money, there is no reason why all investors should prefer immediate dividends. Moreover, shareholders who desire cash can create "homemade dividends" by selling some of their shares.

A9. Share repurchases and dividend payments are substitutes for one another in the sense that both are methods of distributing cash to shareholders. In a perfect capital market, share repurchases and dividend payments are perfect substitutes; both will leave the shareholders' total wealth unchanged.

A10. Assuming the stock is trading rights-on, by Equation (17.2),

$$R = (S - P_R)/(N - 1) = (55\text{-}47)/4 = \$2.00$$

B1. Dividend policy irrelevance is based on the concept of value conservation and value additivity. In a perfect capital market environment, a firm that holds its capital investment program and capital structure constant cannot change its value by choosing a particular dividend policy. Three capital market imperfections that might cause dividend policy to affect the value of the firm are (1) taxes, (2) transaction costs, and (3) asymmetric information. Taxes might affect firm value through dividend policy when investors pay higher taxes on dividend income than on capital gains. If all investors faced this tax differential, investors would be willing to pay a premium for the stock of low-cash-dividend stocks, and the firm could increase its value by retaining and reinvesting earnings. Transaction costs might affect firm value through dividend policy due to the costs of floating a new issue of common stock. In this case, firm value is increased by retaining earnings rather than issuing stock to finance a dividend distribution because the flotation costs of issuing new stock is avoided. Asymmetric information might affect firm value through dividend policy when managers possess more accurate information than investors about the firm's future earnings prospects. In this case, if the firm has better earnings prospects than investors believe and if investors believe that dividend increases signal an increase in expected future earnings, then the firm's value could be increased by increasing dividends.

B2. The cost of floating a new stock issue might cause a firm to choose a lower payout dividend policy than it would otherwise. Every dollar that the firm pays out in dividends is unavailable for positive-NPV investment opportunities. If the firm has paid out a large portion of its earnings, it may have to raise funds externally in order to fund its positive-NPV investment opportunities, and it may be necessary to issue new equity to maintain its target capital structure. As a result, the firm may choose a lower payout dividend policy in order to avoid the flotation costs of a new stock issue.

B3. a. Prior to the dividend, an owner of a share of stock held an asset worth \$100. After the \$20 dividend, the share of stock is worth \$80 and the share's owner has \$20 in cash. Thus, the shareholder's wealth is unchanged at \$100 per share owned. In the aggregate after the dividend, old shareholders have shares worth \$640,000 (= 8000×80) and cash of \$160,000 (= 8000×20). The combined total of shareholders' stock and cash after the dividend is \$800,000 (= 640,000+160,000), exactly what it was prior to the dividend.

To raise the money needed to pay the dividend, the firm must sell 2000 (= 160,000/80) new shares. After paying the dividend and selling the shares, the firm has 10,000 (= 8000 + 2000) shares outstanding. The old shareholders now own 80% (= 8000/10,000) of the newly capitalized firm

b. To raise \$160,000, the existing shareholders must sell 1600 (= 160,000/100) shares. After selling their 1600 shares, the existing shareholders will own 8000 - 1600 = 6400 shares or 80% (= 6400/8000) of the firm. Thus, their percentage ownership of the firm has not changed relative to part a. above. In the aggregate, the existing shareholders own shares worth \$640,000 (= 6400×100) and cash of \$160,000 (= 1600×100), for a total wealth of \$800,000. Exactly what it was prior to the sale of shares and exactly what it is in part a. above.

B4. Using Equation (17.1):

$$DPS_1 - DPS_0 = ADJ \times [POR \times EPS_1 - DPS_0]$$

$$DPS_1 - 1.00 = 0.75 \times [(0.25 \times 8.00) - 1.00] = \$1.75$$

$$DPS_2 - 1.75 = 0.75 \times [(0.25 \times 8.00) - 1.75] = \$1.94$$

$DPS_3 - 1.94 = 0.75\times[(0.25\times8.00) - 1.94] = \$1.99$

$DPS_4 - 1.99 = 0.75\times[(0.25\times8.00) - 1.99] = \$2.00$

B5. We can apply Equation (17.5) by substituting $P_E = P_R - R_P$:

$$R_P = \frac{(1-T_g)\times S + T_g\times B - (P_R - R_P)}{(1-T_g)\times N} = \frac{(1-0.14)\times30 + 0.14\times(17.50) - 23 + R_P}{(1-0.14)\times4} = 1.52616 + \frac{R_P}{3.44}$$

Solving for R_P, we have $R_P = \$2.15$

B6. (1) If the unexpected cut in the dividend rate signals an unexpected decrease in the expected future earnings of the firm, the market price of the firm's stock would drop as investors sell the stock in response to that negative signal. (2) If the unexpected dividend cut caused the firm's stock to be undesirable to a particular tax-oriented clientele (for example, corporate investors who can claim the 70% dividends received deduction), the market price of the stock would drop in response to that clientele rapidly selling its shares. (3) Income-oriented investors would find it costly to replace actual dividends with "homemade dividends," due to the existence of commission charges, and might sell shares in response to a dividend cut in order to reinvest in other higher yielding shares.

B7. Before the dividend ex-date, the stock is trading with the right to receive the dividend. On the ex-date, the stock begins trading without the right to receive the dividend. Therefore, we would expect the stock price to fall by the amount of the declared dividend. It doesn't matter whether the dividend is paid in cash or additional stock because the dilution of the shareholder's claim in a stock dividend is equal to the value drop due to a cash dividend. That is, a stockholder of record who sells the stock ex-dividend, but before the dividend is paid, still receives the dividend. Since the dividend is valuable, whether it is paid in cash or additional stock, the stockholder who purchases the stock ex-dividend will not get that value, hence the decline in stock price in either case. The difference between a cash and a stock dividend is that the cash dividend transforms the stockholders' claim from (1) a proportionate claim on a firm into (2) cash plus a proportionate claim on the same firm which now has less cash in its cash account. A stock dividend does not involve such a transformation; it simply breaks the claim into smaller divisions.

B8. With the firm's capital budget program and capital structure fixed, dividend policy involves a risk and time-value trade-off between present and future dividends. In a perfect capital market, however, the present value of the entire dividend stream, and therefore the stock price, is unaffected by this trade-off.

B9. To the extent that a dividend reinvestment plan can decrease the transaction costs of a transaction that will take place in any case, it can increase shareholder wealth. The transaction costs might be reduced in at least two ways. First, suppose the firm has a high-dividend-payout policy and the reinvestment plan allows shareholders who would use the cash dividends to purchase more of the firm's common stock to purchase the common stock more cheaply through the reinvestment plan. Those shareholders are better off if the dividend reinvestment plan allows them to reinvest the money more cheaply, perhaps with little or no brokerage commission. This also might make the stock attractive to a new group of investors who prefer capital gains to dividend income. Second, if the reinvestment plan allows the firm to make fewer new equity issues, then all of the shareholders are better off because the firm will have lower issuance costs.

B10. The firm's aggregate value before announcement of the bonus dividend is $20 million (1 million shares at $20 per share). Under the bonus dividend alternative, the firm will distribute $5 per share and the ex-dividend price of the stock will be $15 per share. The shareholders' aggregate wealth is unchanged; they own $15 million worth of shares (1 million shares at $15 per share) and have $5 million in cash, so their aggregate wealth is still $20 million. Under the share repurchase alternative, the firm repurchases 250,000 shares at $20 per share. The aggregate wealth of the shareholders is again unchanged; those who sold their shares have $5 million in cash and those who kept their shares have $15 million worth of stock (750,000 shares at $20 per share) for an aggregate wealth of $20 million. Therefore, the shareholders are indifferent between the bonus dividend and the share repurchase in a perfect capital market.

B11. Since the projects are expected to earn their required return, the reinvestment alternative has a zero NPV. Therefore, as we illustrated in section 6 of Chapter 5 in the International Paper example, shareholders will be indifferent to growth versus income when the growth has a zero NPV.

B12. If a firm maintains a policy of never decreasing the per share dollar amount of the cash dividend, any deviation from that policy would signal a worsening of the firm's earnings prospects. Similarly, if the firm always adheres to that policy, a worsening of the firm's financial and business situation would manifest itself in an increase in the payout ratio. In the former case, the principals can monitor the agents by watching for changes in the dividend rate. In the latter case, they can monitor the agents by watching for changes in the payout ratio.

B13. a. Yes. If the firm had only split its shares, its post-split dividend rate would have been $0.21 per share (= 0.42/2), which is less than $0.25.

b. In a perfect capital market, we would not expect any price change. However, in a market characterized by asymmetric information, we would expect the stock price to rise in price as the change in dividend policy is taken as a positive signal by the market. On average, stock prices do increase in reaction to such changes.

B14. a. The announcement of a share repurchase program should have little effect on the firm's share price. Shareholder value is the same whether the distribution is through a cash dividend or a repurchase of shares, and the earnings per share increase is exactly offset by a drop in the P/E ratio. Price effects that do occur are likely to be the result of signaling. For example, if the repurchase signals a lack of positive-NPV projects, then we might expect a price drop. If the repurchase signals undervalued shares, we might expect a price increase.

b. Tender offer announcements tend to provide a more positive signal since managers are willing to repurchase the shares at a premium over the prevailing market price. As a result, we would expect a price increase upon the announcement of a tender offer share repurchase.

B15. The firm might pay a dividend in spite of a need for expansion capital and the flotation costs associated with a new issue because of the information content of dividends. If the dividend payment was expected based on the firm's past dividend policy, investors are likely to interpret the missed dividend as a negative signal of future earnings prospects rather than as a positive signal of the existence of a positive-NPV investment opportunity. As a result, the share price would drop in response to the missed dividend, and this price decrease might more than offset the flotation cost savings associated with avoiding the new issue.

B16. a. Ignoring tax effects, stockholder wealth is unchanged in either case. However, stockholders would pay $8 million in taxes on the dividend distribution and only $4 million in taxes on the capital gains realized in the repurchase, so shareholder wealth is $4 million higher under the repurchase alternative than under the cash dividend alternative.

b. Cardinal's shareholders would be better off having Cardinal retain and reinvest the cash in short-term financial instruments only if the firm anticipates positive-NPV investment opportunities in the near future. In that case, although the short-term financial securities are otherwise zero-NPV investments, the costs that could be avoided by not having to raise external capital in the near future can make them a positive-NPV investment.

B17. The current stated tax rate on capital gains is higher than the 10% rate described in this problem, so a change to the 10% rate would increase the tax bias in favor of retained earnings. Therefore, average payout ratio of NYSE-traded firms might be expected to drop. However, this drop in payout ratio would probably not be significant because of the tax-timing option and the large proportion of investment that is either tax exempt or tax deferred.

B18. Paying a cash dividend would be a significant committment of cash that would be consistent with, and more meaningful than, the chairman's statement that he expects "very positive" 1996 results. Such a dividend would be a visible signal of management's belief, and the signal would subsequently be either confirmed or disproven.

B19. Optimize your firm's capital structure. The existence of investor clienteles mitigates many of the potential gains associated with dividend policy, so that the choice of a *particular* policy is often less critical than simply maintaining a stable dividend policy. On the other hand, capital structure optimization requires the analysis of a number of factors (such as agency costs, signaling considerations, and bankruptcy costs) that are likely to be difficult to quantify and may have conflicting effects on the capital structure decision. As a result, there is a greater potential for gain from capital structure policy, and therefore you should ask the good fairy to tell you how to optimize your firm's capital structure policy.

C1. a. An increase in the dividend rate might lead to an increase in a firm's share price when that dividend increase signals improved expectations for future earnings.

b. The firm would prefer to pay out the cash in the form of a "special" or "extra" dividend when it did not anticipate that the dividend increase could be sustained. In this case, designating the dividend as a "special" distribution might allow the firm to avoid the price drop we would expect when the dividend went back down to its normal level the next quarter.

c. The renewed interest in cash dividends does not imply that dividends are more valuable to investors because they are more certain than expected future earnings. It simply means that firm's lacking profitable investment opportunities should distribute excess cash to their investors. Otherwise these cash rich and investment poor firm's may become takeover targets for corporate raiders.

C2. As with the traditional view of capital structure, the traditional view of dividend policy is wrong because of the reasoning that leads to its conclusion rather than because the conclusion is incorrect. The traditional view that dividend policy is relevant because investors prefer "dividends in the hand" to risky capital gains is incorrect. The conclusion that dividend policy is relevant results from the interaction of a number of capital market imperfections such as taxes, transaction costs, and asymmetric information.

C3. a. The stock market normally reacts unfavorably to the announcement of a dividend reduction.

b. If a dividend reduction signals that future earnings are expected to be insufficient to sustain dividends at the previous level, then the dividend reduction provides a negative signal regarding the firm's future earnings prospects.

c. No, this is not inconsistent. The price increase upon announcement of the dividend reduction is only inconsistent if the announcement contained no new information. The price increase can be the result of market participants having *expected* a larger decrease in the dividend rate and being surprised that the reduction was only $0.25. Also, the change in share price must be measured relative to the market as a whole. For example, the change in share price to $62 from $61 represents a 1.639% increase. If the Standard & Poor's 500 Index, a useful proxy for share prices generally, rose 2.639% that day and the shares have a beta of 1.0, then the share price actually fell by 1% (= 2.639 - 1.639) in response to the dividend reduction announcement when measured relative to the market.

d. If the reduction is less than expected. Another possibility is that the firm released other positive information simultaneous with the dividend announcement. For example, the firm might announce a major restructuring or a major new investment program that will be partially financed with the increased retentions. The positive signal associated with these other announcements might more than compensate for the dividend decrease.

C4. Firm's are reluctant to cut their dividend rate because of the signaling content of dividends. A cut in dividend rate, even when it is accompanied by the announcement of a proposed capital investment, generally results in a significant stock price decrease. An unexpected dividend cut or a dividend increase that is less than expected is viewed by the market as a signal that the firm's future earnings prospects are worse than previously expected. As a result, firms are reluctant to cut their dividend or to increase dividends above a sustainable rate.

C5. A stock that will *absolutely never* pay a dividend is worthless. As a result, prohibiting utilities from ever paying cash dividends would result in investors being unwilling to hold utility stock at any positive price. In a sense, the utilities' cost of capital would become infinite because no investor would be willing to pay anything for their shares.

C6. A common method of maintaining a relatively stable capital structure over time is to use retained earnings (equity) and debt issues in the chosen proportion to finance new projects. If the earnings are distributed as dividends, then they are unavailable for financing investment projects and the firm may be forced to forgo a profitable investment opportunity rather than take on additional partners (issue new shares). In that case, the dividend policy would give rise to the opportunity cost of the forgone investment if the capital structure is held constant.

C7. a. Consider a shareholder who owns n shares. The total value of his holding before the put rights offering is $n \times P_R$. After exercising his put rights to sell n/N shares, he will have $n \times S/N$ in cash and $(n - n/N)$ shares worth P_E each. For the value of his holding to be unchanged,

$$(n - n/N) \times P_E + n \times S/N = n \times P_R$$

Multiplying through by N/n and solving for P_E gives

$$P_E = (N \times P_R - S)/(N - 1)$$

b. Equation (17.3) gives $R_P = P_R - P_E$. Therefore, substituting the expression from part a into this expression for R_P gives,

$$R_P = P_R - (N \times P_R - S)/(N - 1) = [P_R \times (N - 1) - N \times P_R + S]/[N - 1]$$

$$= (N \times P_R - P_R - N \times P_R + S)/(N - 1)$$

$$= (S - P_R)/(N - 1)$$

c. Using the reasoning applied in part a and solving instead for the rights-on value of the stock,

$$P_R = [1 - (1/N)] \times P_E + S/N = [(N - 1) \times P_E + S]/N$$

Equation (17.3) gives $R_P = P_R - P_E$. Therefore,

$$R_P = [(N - 1) \times P_E + S]/N - P_E = [(N - 1) \times P_E + S - N \times P_E]/N = (S - P_E)/N$$

C9. a. Let m denote the total number of shares prior to the put rights offering. After the announcement but before the ex-date, the total value of the firm is $m \times P_R$. At exercise, the number of shares repurchased will be m/N, so there will be $m - m/N$ shares outstanding, and the firm will pay $(m/N) \times S$ to repurchase shares. Therefore, after exercise, firm value will be $[m - (m/N)] \times P_E$ and that must equal the value before minus the cost, or

$$[m - (m/N)] \times P_E = m \times P_R - (m/N) \times S$$

Solving this expression for P_E gives

$$P_E = (N \times P_R - S)/(N - 1)$$

b. A shareholder can receive S for her share by purchasing N put rights, exercising the options, and paying taxes of $T_g \times (S - B - N \times R_P)$. Thus, ex-rights her share will be worth

$$P_E = S - N \times R_P - T_g \times (S - B - N \times R_P) = (1 - T_g) \times S + T_g \times B - (1 - T_g) \times N \times R_P$$

c. Solving the right-hand expression in part b we get,

$$R_P = [(1 - T_g) \times S + T_g \times B - P_E]/[(1 - T_g) \times N]$$

d. Substituting Equation (17.3) into the expression in part c and solving for R_P, we get

$$R_P = [(1 - T_g) \times S + T_g \times B - P_R]/[(1 - T_g) \times N - 1]$$

CHAPTER 18 -- MANAGING DIVIDEND POLICY

A1. April 10 is the declaration date; April 24 is the ex-dividend date; April 27 is the record date; and May 5 is the payment date.

A2. The most common pattern of dividend policy is (b): a stable payment rate per share where dividend payments are more stable than earnings.

A3. Shareholders of record as of the close of business on the **record date** will receive a check for the dividends declared on the shares they own as of the record date.

A4. The three-step approach to the dividend decision is (1) estimate future residual funds based on projected levels of capital expenditures and operating cash flows; (2) determine an appropriate target payout ratio based on an analysis of comparable firms and the analysis in step one; and (3) set the quarterly dividend rate at a sustainable level taking into consideration the signaling content of dividend changes.

A5. True.

A6. The primary difference between a stock dividend and a stock split is the transaction's impact on the firm's accounting records. With a stock dividend, the fair market value of the distributed shares is transferred from the retained earnings account to the "paid-in-capital" and "capital contributed in excess of par value" accounts on the firm's balance sheet. With a stock split, the par value of the shares is changed but there is no transfer among balance sheet equity accounts.

A7. The major advantage of a share repurchase over a cash dividend is the favorable tax treatment of the repurchase; taxable individual shareholders generally suffer a lesser tax liability and hence realize greater wealth if the corporation repurchases shares rather than distributing a cash dividend.

A8. A large share repurchase might have an impact on a firm's debt rating because the higher leverage resulting from the repurchase might decrease the firm's financial flexibility and thereby contribute to a downgrading. This is only likely to be an important factor if the repurchase program is large relative to the firm's capitalization and the rating is already in jeopardy.

A9. False. Two factors that could preclude a firm from paying a cash common dividend are: (1) the firm's outstanding debt may have restrictive covenants limiting dividend payments that prevent the firm from paying out the full amount it would like to distribute and (2) the firm may have preferred dividends in arrears that must be paid before any distribution can be made to common shareholders.

A10. True.

A11. In an open market purchase program, a firm repurchases its shares on the open market at prevailing prices, and the purchases are subject to SEC regulations aimed at minimizing the impact of the repurchase program on the firm's share price. With a tender offer, the firm offers to purchase a stated number of shares at a fixed price or through a Dutch auction. The tender offer method has the advantage of providing all shareholders an equal opportunity to sell their shares back to the firm, and it generally provides a more positive signal to the market than an open market repurchase. The disadvantages of a tender offer are the higher transaction costs, the premium paid to induce tendering, and the risk of oversubscription.

A12. With a Dutch auction tender offer, individual shareholders specify the minimum price (within a range specified by the firm) at which they are willing to tender shares and the amount they are willing to tender at that price. If less than the minimum number of shares sought are tendered, the firm must pay the maximum price in its specified range; otherwise, the firm selects a price within the specified range and pays that price for all shares tendered at or below that price. With a fixed-price tender offer, the firm pays a single pre-specified price for all shares tendered. As a result, the Dutch auction method may lead to a lower cost of repurchase than the fixed-price method.

A13. The principal advantages of the transferable put rights method of share repurchase relative to the fixed-price tender offer method are: (1) put rights allow those shareholders who do not want to tender to receive value for their rights by selling rights to other investors; (2) put rights are more tax efficient than a fixed-price tender offer; and (3) there is no risk of oversubscription because the number of put rights issued limits the number of shares that can be tendered.

B1. a. Rearranging the dividend growth model gives

$$r = D_1/P_0 + g = 2/25 + 0.10 = 18\%$$

b. In a perfect capital market environment, shareholder value is not affected by dividend policy, so the share price will remain \$25. In a perfect capital market, the value of the firm's stock is the same whether earnings are retained or paid out as dividends financed with new equity offerings.

B2. A firm might find it advantageous to substitute a stock dividend for an increase in its cash dividend when it needs to conserve cash but still wants to convey positive information regarding future earnings prospects. However, if investors interpret the substitution as an expression of a lack of confidence in the firm's (possibly improved) future prospects, the positive effect of the dividend signal will be diluted, or possibly eliminated, with little or no change in share price occurring.

B3. Consider a firm with 1 million shares outstanding at a current market value of \$20 per share. The firm has \$2 million to distribute either as a cash dividend (\$2/share) or by repurchasing 100,000 shares. In the case of the cash dividend, the share price will drop on the ex-dividend date to \$18 per share and shareholders will still have a total wealth of \$20 per share (\$2 in cash plus a share worth \$18). In the case of the repurchase, the new share value will be [1,000,000×(\$20) - \$2,000,000]/[1,000,000 - 100,000] = \$20. In this case, shareholders who sell will have \$20 cash and those who don't will still have a share worth \$20. The net result is that shareholder wealth is the same with the cash dividend as with the share repurchase in a perfect capital market.

B4. Consider a firm with 1 million shares outstanding at a current market value of \$20 per share and debt outstanding at a current market value of \$10 million. Then the firm's initial debt ratio is $L = 1/3$ (= 10/(10 + 20)). If the firm distributes \$2 million by repurchasing 100,000 shares, the total number of shares outstanding will decrease to 900,000 and the total market value of equity will become \$18 million. The amount of debt outstanding is unchanged by the transaction, and the new debt ratio is $L = 10/(10 + 18) = 0.357$.

B5. a. The maximum overall payout ratio is the total discretionary cash flow divided by the total earnings, which is 215/635 = 33.86%.

b. To pay out all discretionary cash flow in years 1 through 5, the total dividend over the 5 years must be \$10.75. The following recommended dividend policy maintains stable growth in the per share dividend, makes what appear to be sustainable increases, and pays out most of the discretionary cash flow:

Year	Dividend	Payout Ratio
1	\$1.70	34.0%
2	1.90	30.4
3	2.10	28.0
4	2.35	39.2
5	2.60	37.1

B6. a. The total 1997 dividend under the new policy will be $4\times(0.40) + 0.3\times(7.50 - 6.00) = \$2.05$, and this represents 27.3% (= 2.05/7.50) of its 1997 earnings.

b. Alcoa's new dividend policy would be appropriate for firms in cyclical businesses. This policy allows such firms to pay out a stable percentage of their earnings without requiring borrowing during business troughs and excess cash buildup during business peaks. This policy also allows the firms to preserve the integrity of dividend changes as a signaling device.

B7. a. $APY = [1 + (0.12/4)]^4 - 1 = 12.55\%$

b. In this case, the stock is a perpetuity of quarterly \$0.25 payments with a required return of 3% per quarter, compounded quarterly. As a result, the value of the stock is $P_0 = \$8.33$ (= 0.25/.03).

c. With an annual dividend, the stock is a perpetuity of annual \$1.00 payments with an APY required return of 12.55%. In this case the value of the stock is $P_0 = \$7.97$ (= 1.00/0.1255). The value is different with the annual dividend because of the time value of money. The quarterly dividends of \$0.25 each can be invested as they are received and are worth more than the single dollar received at the end of the year.

d. Pattern 3 provides a long-term payout ratio of 1/3 without requiring any dividend decreases.

B8. Typically, a small and growing firm will not pay cash dividends to its common stockholders, but instead reinvest all its earnings in the rapidly expanding business. At some point, the firm decides that it is mature enough to pay regular cash dividends and adopts a policy of paying a small stable per share dividend. As earnings increase, the firm may make dividend increases that it believes are sustainable. This is the policy the firm will maintain for most of its life. If the firm begins to decline financially, it will reluctantly cut, and perhaps eventually eliminate, common dividends as it approaches its demise.

B9. a. The special dividend announcement suggests that Allegis did not have profitable investment opportunities in which to invest the proceeds from the sale of its subsidiaries.

b. By announcing a special dividend, investors will not expect the dividend increase to be sustained so there is no negative information when the dividend returns to its normal level.

c. The shareholder's preference for a share repurchase rather than a special dividend seems to indicate that the shareholder is a high-marginal-tax-rate investor who pays a higher effective tax rate on dividends than on capital gains.

B10. a. With $10 million to distribute among 5 million shares, the special dividend would be $2.00 per share, and the share price would be expected to drop to $18.

b. If the firm repurchases shares at the $20 market price, it can repurchase 500,000 shares.

c. Under the special dividend alternative, EPS would be $2.00, and under the repurchase alternative, EPS would increase to $2.22 since the $10 million in total earnings ($2.00 per share × 5 million shares) would be distributed among 4.5 million shares. An efficient market would not react to the change in EPS because it would realize that the increase in EPS is perfectly offset by a decrease in the P/E ratio.

d. In a perfect capital market, the cash distribution causes the share price to drop to $18, so the shareholders' total wealth is left unchanged ($2 in cash and a share worth $18 for a total value of $20 per original share). Following the share repurchase, each share would be worth [5,000,000×(20) - 10,000,000] /[5,000,000 - 500,000] = $20 so, again, the shareholders' total wealth is left unchanged. Those who sell their shares have cash of $20 and those who don't have a share worth $20. In a perfect capital market, a share repurchase is equivalent to a cash dividend distribution.

B11. a. After the special dividend, shareholders will have a share worth $18 and a $1.00 after-tax cash dividend, so aggregate shareholder wealth is $95 million (5 million shares x $19 per share). With the repurchase, the shares are sold for $20 each, $10 of which is a capital gain. Shareholders are taxed at 25% on the $5,000,000 aggregate capital gain so aggregate shareholder wealth is $98.75 million (= 100 million - 0.25×[5 million]).

b. Because the shares would otherwise be sold at the end of 5 years, the share repurchase accelerates the capital gain tax liability. The present value increase in the capital gain tax liability is:

$$1{,}250{,}000 - 1{,}250{,}000/(1.12)^5 = \$540{,}716.43$$

So the aggregate wealth with the repurchase would be $99.46 million versus $95 million for the special dividend.

c. When dividends are taxed at a higher effective rate than capital gains, shareholders will prefer a share repurchase to a special dividend distribution from a purely tax standpoint.

B12. a. Substituting a stock dividend for half of the usual cash dividend will not solve the problem. The stock dividend is more expensive administratively and, unless the dividend rate is cut later, the stock dividend will mean an even higher total dividend payment in the future as more shares of stock will be outstanding. In addition, if investors feel that the substitution is due to the rate increase denial and accompanying financial difficulties, as is likely to be the case, they will react negatively to the substitution.

b. Substituting debentures for a portion of the usual cash dividend also fails to solve the problem. This would be viewed by investors as being essentially equivalent to issuing debt to finance the usual dividend payment. In this case, the firm will incur future interest payments to service the debenture and will still have the inherent problem of insufficient earnings. As with the stock dividend, investors will not be fooled by this "dividend substitution".

B13. a. Prior to the buyback, aggregate stockholder wealth is $300 million, and $35 million of this will be distributed to Karl I. Can. After the buyback, aggregate stockholder wealth is $265 million and there are 9 million shares remaining. The firm's share price immediately following the buyback is $29.44 per share (= 265 million/9 million shares).

b. The payment of greenmail causes the value of the remaining shareholders' stock to drop by $5 million. The buyback results in a $5 million wealth transfer from the remaining shareholders to Karl I. Can.

B14. a. Because she owns 5 million shares and will be receiving a $3 per share premium, the repurchase would increase the minority shareholder's wealth by $15 million.

b. Prior to the repurchase, aggregate stockholder wealth is $600 million (50 million shares at $12 per share), and $75 million of this will be distributed to the minority shareholder. After the repurchase, aggregate stockholder wealth is $525 million, and there are 45 million shares remaining. Easy Mark's share price would drop to $11.67 per share (= 525 million/45 million shares) following the repurchase.

c. The aggregate wealth of the other shareholders would drop by $15 million (= [12 - 525/45]×45).

d. Greenmail represents the expropriation of wealth from the other stockholders for the benefit of the greenmailer.

B15. Because the cost of servicing a small account is nearly the same as the cost of servicing a large account, the firm may find it advantageous to repurchase small shareholdings that are outstanding. In addition, the opportunity to receive a premium price and to avoid the disproportionately high brokerage commissions on odd-lot transactions makes it advantageous for the small account owner to tender her shares.

B16. Yes. If shares are sold between the ex-dividend date and the payment date, the person who sold the shares will receive the dividend on the payment date even though he no longer owns the shares, and the person who purchased the shares will not receive the dividend on the payment date even though he does own the shares.

B17. a. A 50% stock dividend would require the distribution of 2.5 million shares worth $44 per share. This amount ($110 million) is transferred out of retained earnings. Because the number of shares outstanding is increased by 2.5 million shares and the par value is unchanged, the paid-in capital account is increased by $10 million. The accounts are balanced by transferring the remaining $100 million into the capital contributed in excess of par value account. These changes are illustrated below:

Paid-in capital ($4 par value; 7.5 M shares)	$ 30,000,000
Capital contributed in excess of par value	130,000,000
Retained earnings	-60,000,000
Common stockholders' equity	$100,000,000

b. With a 3-for-2 stock split, the par value of the shares decreases to $\$2.66^2/_3$ per share, and the common equity account balances remain unchanged:

Paid-in capital ($\$2.66^2/_3$ par value; 7.5 M shares)	$ 20,000,000
Capital contributed in excess of par value	30,000,000
Retained earnings	50,000,000
Common stockholders' equity	$100,000,000

c. Neither the stock dividend nor the stock split alters the financial position of the firm but they might nevertheless have favorable signaling effects. Empirical evidence suggests that the announcement of a stock split generally has a favorable impact on share price. Because a 50% stock dividend is likely to have the same informational content as a 3-for-2 stock split, the market impact of the stock dividend announcement is unlikely to be significantly different from the market impact of the stock split announcement.

B18. a. When dividend increases are used as signals of an improving corporate outlook, we expect the share price to rise after the announcement of the increase. However, in this instance conflicting signals are being sent. Kerkorian is pressuring Chrysler to increase the dividend and may be forcing management to behave in manner inconsistent with what management perceives to in the best interest of shareholders. In addition, reducing Chrysler's "excess cash" might lessen its attractiveness as a takeover candidate. Thus, it would not be surprising to see Chrysler's share price fall slightly.

b. The choice in this case involves the trade-off between the lower cost, open market repurchase and the stronger positive signal sent by a tender offer. Given that Chrysler is responding to outside pressure and has also raised its cash dividend, it may as well choose the lower cost alternative--the open market repurchase.

c. Corporate taxpayers pay lower taxes on dividends than on capital gains due to the 70% dividend exclusion. Individual taxpayers pay lower taxes on capital gains than dividends. Since Chrysler is a widely held firm, it is partially satisfying both corporate and individual shareholders by increasing its dividend and repurchasing shares.

B19. The Dutch auction procedure only allows shareholders to select the share repurchase price within a range specified by the firm. If the firm would have been willing to pay the maximum for all the shares, it may benefit by getting to pay less for at least some of them.

B20. a Under Alternative C, the total per share dividends for 1997-2001 are $11.50. The firm has 10 million shares outstanding, so the firm's aggregate dividend is $115 million (= 11.50×10 million).

b. In the table below, Earnings per share and Residual funds are from Table 18-4. The Dividend per share is stated in the problem. The Payout ratio is calculated as the Dividend per share divided by the Earnings per share. Dividend requirements are calculated as the Dividend per share multiplied by the 10 million shares outstanding. The Surplus (deficit) is calculated as Residual Funds minus Dividend requirements.

Alternative C	1997	1998	1999	2000	2001	Totals
Earnings per share	$ 5.00	$ 6.00	$ 7.20	$ 8.50	$ 10.00	
Dividend per share	$ 1.64	$ 1.84	$ 2.12	$ 2.56	$ 3.34	$ 11.5
Payout ratio	33%	31%	29%	30%	33%	
Residual funds (millions)	$ 17.0	$ 20.0	$ 14.0	$ 32.0	$ 32.0	$ 115
Dividend requirements (millions)	16.4	18.4	21.2	25.6	33.4	115
Surplus (deficit) (millions)	$ 0.6	$ 1.6	$ (7.2)	$ 6.4	$ (1.4)	0.0

c. Under all three alternatives, Major pays out all estimated residual funds to shareholders. Alternative C, like Alternative B, provides for increases in each year and avoids the large increases in 2000 and 2001 called for under Alternative A. In addition, Alternative C's payout ratio is also closer to the long run target ratio than is A's.

Alternative C offers a more stable payout ratio than Alternative B (a range 29%-33% versus a range of 28-34% respectively). However, the more stable payout ratio comes at the expense of a less stable surplus/deficit for Alternative C. Finally, Alternative C calls for smaller increases in early years and larger increases in later years than Alternative B.

d. The choice is between Alternatives B and C. Alternative B calls for a $0.04 larger increase in 1997 than does Alternative C. Since the treasurer is projecting higher earnings than are expected by analysts and it appears that the higher dividend rate is sustainable, the stronger signal sent by the larger dividend increase is probably warranted.

C1. In an efficient capital market, the share price will already have dropped as a result of the poor economic conditions and the price will already reflect the possibility that the dividend will be cut. While denial on the part of the board might change the share price movement in the very near-term, it can substantially damage the firm's ability to credibly signal information in the future.

C2. a. Firm A is sending the more credible signal. It supports its statement that the stock is undervalued by its willingness to pay a premium over the current market price.

b. If the two firms are, in fact, identical except for their repurchase plans, then you should buy the shares of Firm B. While the stock of Firm A may exhibit an immediate positive reaction to the tender repurchase announcement, by the end of the three years, Firm B stock will also have appreciated to its true value. Firm B will have avoided the premium cost of the tender offer, thereby retaining more wealth for the remaining shareholders.

C3. a. The firm could offer a swap of either preferred stock for common stock, or debt for common stock, depending on the tax status of the firm and the tax positions of its income-oriented investors. The new issue of preferred stock or debt would be designed to preserve the pre-dividend-cut income level of the shareholders who swap--by providing $4 of annual income from the new securities for each common share swapped--and could be convertible into, or redeemable for, the issuer's common stock within, say, a few years.

b. The plan to eliminate dividends should be announced before any dividend cuts occur, along with the reasons for the cut and a description of the increased profits that are expected from the shift in business strategy.

c. Although the dividend elimination announcement will likely have a negative stock price reaction, that reaction should be tempered by a positive reaction to the proposed (and profitable) new strategy. In addition, since investors anticipate the elimination over a longer time period, a sudden, sharp drop due to selling by the income-oriented investors should be avoided.

C4. With the separation between the firm's ownership and its control, the firm's managers know more about the daily operations of the firm and about its prospects for the future than do the shareholders. As a result, the managers need a way to convey credibly their information to the firm's shareholders. Paying out a fixed portion of its net cash inflow conveys information regarding only the current period's cash inflow; it does not provide a means of signaling future prospects. With a smoothed dividend policy and a reluctance to cut dividends, shareholders expect a particular dividend level and they know that dividend levels are only increased if management expects to be able to sustain that increase. Because of this, changes in the dividend level provide an accurate signal of future prospects.

C5. Unless there is a differential in the tax treatment of dividend income and capital gains, the Polish shareholders should be indifferent between cash dividends and appreciation in the value of their stock; both of these returns will act to improve their standard of living. In addition, because capital is a scarce resource, the government should consider preferential tax treatment for capital gains. In this case, Polish firms will be able to utilize the relatively inexpensive internal equity at the same time that investors prefer low-dividend stocks.

CHAPTER 19 -- CASH AND WORKING CAPITAL MANAGEMENT

A1. Three alternative approaches to working capital management are the maturity-matching approach, conservative approach, or aggressive approach. In the maturity matching approach, the firm matches the maturities of its assets and liabilities. Seasonal variations in current assets are financed with current liabilities of the same maturity; however, any permanent component of current assets, such as some portion of inventories and receivables, is financed with long-term capital.

The conservative approach uses more long-term and less short-term financing than the maturity matching approach. Long-term financing is used to finance all the firm's long-term assets, all of its permanent current assets, and some of its temporary current assets. This guards against the risks of a credit shutoff or an increase in the cost of funds and builds in a margin of safety.

The aggressive approach uses less long-term and more short-term financing. Since the cost of long-term funds generally is greater than the cost of short-term funds, financing working capital with relatively more short-term funds raises expected profitability and increases risk.

A2. The cash conversion cycle is the length of time between payment of cash for inventory and receipt of cash from accounts receivable. The cash conversion cycle is equal to the inventory conversion period plus the receivables collection period minus the payables deferral period. This cycle is important to working capital management because the longer the cash conversion cycle, the greater the need for increased levels of working capital financing.

A3. a. By Equation (19.2)
Inventory conversion period = $365 / (10.0 / 1.0) = 36.5$ days

b. By Equation (19.3)
Receivables collection period = $365 / (25 / 0.8) = 11.7$ days

c. By Equation (19.3)
Payables deferral period = $365 \times (0.4 + 0.15) / (10.0 + 1.5) = 17.5$ days

d. By Equation (19.1)
Cash conversion cycle = $36.5 + 11.7 - 17.5 = 30.7$ days

A4. The three basic motives for holding cash are: (1) the transactions motive--the need to hold cash to make payments in the ordinary operation of the business; (2) the precautionary motive--the need to hold cash to meet unexpected contingencies; and (3) the speculative motive--the desire to hold cash in order to be able to take advantage of profitable opportunities that may arise unexpectedly and require an immediate use of cash.

A5. Float is the difference between the actual checking account balance at the bank and the balance on the firm's books. It arises from the delay between the time a check is written and the time the funds are actually taken out of the checking account. The two basic types of float are collection float and disbursement float.

A6. Disbursement float = 20,000 - 18,000 = \$2000

A7. Types of marketable securities: (1) U.S. Treasury securities are issued by the federal government offer the lowest risk and greatest liquidity; (2) U.S. federal agency securities are issued by agencies of the federal government, are nearly as liquid as Treasury securities, and offer slightly higher risk and return; (3) negotiable certificates of deposit are time deposits issued by domestic or foreign commercial banks that can be sold to a third party; (4) short-term tax-exempt municipal securities are issued by state and local governments, often in anticipation of cash receipts; (5) bankers' acceptances are drafts that a commercial bank has accepted; (6) issues of commercial paper are unsecured promissory notes issued by large, creditworthy corporations; and (7) money market preferred shares are securities which have a floating dividend rate that is reset frequently to reflect current market rates.

A8. Cost = Fixed cost + Float costs

Cost of check = 1.00 + (250,000 × 0.03 × (5/365)) = $102.74

Cost of wire transfer = 12.00 + 0 = $12.00

A9. The following are five methods firms use to manage their float: (1) wire transfers are more expensive than writing a paper check, and the float is reduced since the transfer is immediate; (2) zero balance accounts are special disbursement accounts which maintain a zero balance until funds are needed, at which time the funds are automatically transferred into the account; (3) controlled disbursing involves the use of disbursing accounts at several banks in addition to the master account at the firm's lead bank (as funds are needed, the banks notify the firm, and funds are wired to the bank; the banks also notify the firm about excess funds and wire them out) (4) centralized processing of payables involves prompt bill paying of bills by the cash manager so that the firm avoids late penalties and cash balances are reduced the managers knows when the bills are due, and ensures the funds are available and the bills are paid on time) and (5) lockboxes are post office boxes to which firms direct their incoming checks so that float is reduced by shortened mail times, processing times, and availability times.

A10. By Equation (19.9)

APR = (1.5%/(100% - 1.5%)) × (365/(50 - 15) = 15.88%

APR = (1.5%/(100% - 1.5%)) × (365/(75 - 15) = 9.26%

A11. Three types of short-term bank loans are : (1) a transaction loan is made for a specific purpose; (2) a line of credit is an arrangement between a bank and a customer concerning the maximum loan balance the bank will permit the borrower at any one time; and (3) a revolving credit agreement represents a legal commitment to lend up to a specified maximum amount during a specified time period.

A12. The major advantage of using EDI (electronic data interchange) is that it increases efficiency by allowing the firm a much higher degree of control over transactions at a lower cost and with fewer errors.

B1. By Equation (19.2): Inventory conversion period = 365/6 = 60.8 days

By Equation (19.3): Receivables collection period = 365/10 = 36.5 days

By Equation (19.3): Payables deferral period = 365/12 = 30.4 days

By Equation (19.1): Cash conversion cycle = 60.8 + 36.5 - 30.4 = 66.9 days

B2. a. The amount of cash required, T, is \$2 million per month; the opportunity cost of holding cash, I, is 9% per annum or 0.75% (= 9/12) per month; and the fixed cost, b, is \$300. The optimal transaction size, C, is the amount of cash the firm would raise each time it sells securities. Using Equation (19.6):

$$C = \left[\frac{2 \times b \times T}{i}\right]^{1/2} = \left[\frac{2 \times (300) \times (2M)}{0.0075}\right]^{1/2} = \$400{,}000$$

b. The firm should sell securities T/C = \$2 million/\$400,000 per transaction = 5 times per month

c. Average cash balance = $C/2$ = 400,000/2 = \$200,000

d. Annual opportunity cost of funds = $i \times C/2$ = 0.09 x 400,000/2 = \$18,000

Annual transactions costs = $b \times T/C$ = 300.00 × 5 × 12 = \$18,000

Total annual cost = 18,000 + 18,000 = \$36,000

B3. a. Optimal deposit size, by applying Equation (19.6):

$$C = \left[\frac{2 \times b \times T}{i}\right]^{1/2} = \left[\frac{2 \times (200) \times (4M)}{0.04}\right]^{1/2} = \$200{,}000$$

Annual opportunity cost of funds = $i \times C/2$ = 0.04 × 200,000/2 = \$4000

Annual transactions costs = $b \times T/C$ = 200 × 4M/200,000 = \$4000

Total annual cost = 4000 + 4000 = \$8000

b. Annual opportunity cost of funds = $i \times C/2$ = 0.08 × 200,000/2 = \$8000

Annual transactions costs = $b \times T/C$ = 200 × 4M/200,000 = \$4000

Total annual cost = 8000 + 4000 = \$12,000

c. By equation (19.6):

$$C = \left[\frac{2 \times b \times T}{i}\right]^{1/2} = \left[\frac{2 \times (200) \times (4M)}{0.08}\right]^{1/2} = \$141{,}421$$

Annual opportunity cost of funds = $i \times C/2$ = 0.08 × 141,421/2 = \$5657

Annual transactions costs = $b \times T/C$ = 200 × 4M/141,421 = \$5657

Total annual cost = 5657 + 5657 = \$11,314

B4. a. Equation (19.7) gives Z:

$$Z = \left[\frac{3 \times b \times \sigma^2}{4 \times i}\right]^{1/3} = \left[\frac{3 \times (300) \times (10{,}000{,}000{,}000)}{4 \times (0.09/365)}\right]^{1/3} = \$208{,}967$$

The return point is given by:

$RP = LCL + Z = 0 + 208{,}967 = \$208{,}967$

The upper control limit is given by:

$UCL = LCL + 3 \times Z = \$626{,}901$

b. When the cash balance reaches $626,901, the firm should purchase $417,934 (626,901 - 208,967) of marketable securities, and when the cash balance reaches zero, the firm should convert $208,967 worth of marketable securities to cash.

c. Average cash balance $= LCL + (4/3) \times Z = 0 + (4/3) \times 208{,}967 = \$278{,}623$

B5. Cost of skipping the discount and paying at the end of the net period using APR by equation (19.9):

a. $APR = (1\%/(100\% - 1\%)) \times (365/(30 - 10) = 18.43\%$

b. $APR = (6\%/(100\% - 6\%)) \times (365/(70 - 10) = 38.83\%$

c. $APR = (2\%/(100\% - 2\%)) \times (365/(45 - 15) = 24.83\%$

Cost of skipping the discount and paying at the end of the net period using APY by equation (19.10):

a.

$$APY = -1 + \left[1 + \frac{\text{discount\%}}{100\% - \text{discount\%}}\right]^{365/\text{total period - discount period}} = -1 + \left[1 + \frac{1}{99}\right]^{365/20} = 20.13\%$$

b.

$$APY = -1 + \left[1 + \frac{\text{discount\%}}{100\% - \text{discount\%}}\right]^{365/\text{total period - discount period}} = -1 + \left[1 + \frac{6}{94}\right]^{365/60} = 45.70\%$$

c.

$$APY = -1 + \left[1 + \frac{\text{discount\%}}{100\% - \text{discount\%}}\right]^{365/\text{total period - discount period}} = -1 + \left[1 + \frac{2}{98}\right]^{365/30} = 27.86\%$$

B6. a. Cost of skipping the discount and paying on day 40, by equation (19.9):

APR = (2%/(100% - 2%)) × (365/(40 - 10) = 24.83%

$$APY = -1 + \left[1 + \frac{\text{discount\%}}{100\% - \text{discount\%}}\right]^{365/\text{total period - discount period}} = -1 + \left[1 + \frac{2}{98}\right]^{365/30} = 27.86\%$$

b. APR = (2%/(100% - 2%)) × (365/(55 - 10) = 16.55%

$$APY = -1 + \left[1 + \frac{\text{discount\%}}{100\% - \text{discount\%}}\right]^{365/\text{total period - discount period}} = -1 + \left[1 + \frac{2}{98}\right]^{365/45} = 17.81\%$$

B7. By equation (19.8):

Monday: Disbursement float = Available float - Book balance = $100

Tuesday: Disbursement float = $300

Wednesday: Disbursement float = $500

Thursday: Disbursement float = $300

Friday: Disbursement float = $0

B8.

Day	Checks written	Deposits	Book balance	Available balance	Disburse-ment float	Availability float	Total float
1	0	0	10,000	10,000	0	0	0
2	500	1000	10,500	10,000	500	- 1000	- 500
3	800	0	9700	11,000	1300	0	1300
4	800	0	8900	10,500	1600	0	1600
5	0	0	8900	9700	800	0	800
6	0	2000	10,900	8900	0	- 2000	-2000
7	900	0	10,000	10,900	900	0	900
8	0	0	10,000	10,900	900	0	900
9	0	0	10,000	10,000	0	0	0
10	0	0	10,000	10,000	0	0	0

B9. In the calculator solutions below, set PMT = 0.

a. PV = 100 FV = - 115 $n = 1$ solve for r = APY = 15.00%

b. PV = 85 FV = - 100 $n = 1$ solve for r = APY = 17.65%

c. PV = 90 FV = - 105 $n = 1$ solve for r = APY = 16.67%

d. PV = 75 FV = - 90 $n = 1$ solve for r = APY = 20.00%

B10. In the calculator solutions below, set PMT = 0.

a. PV = 100 FV = - 103.75 $n = 0.25$ solve for r = APY = 15.87%

b. PV = 96.25 FV = - 100 $n = 0.25$ solve for r = APY = 16.52%

c. PV = 90 FV = - 93.75 $n = 0.25$ solve for r = APY = 17.74%

d. PV = 81.25 FV = -85.00 $n = 0.25$ solve for r = APY = 19.78%

B11. Lending rate = prime + 1 % = 9% + 1% = 10%

APR = 11.11%

B12. a. APR = 37,500/250,000 = 15%

APY = (287,500/250,000) - 1 = 15%

b. APR = 27,500/200,000 = 13.75%

APY = (227,500/200,000) - 1 = 13.75%

c. APR = 35,000/215,000 = 16.28%

APY = (250,000 \ 215,000) - 1 = 16.28%

Alternative (b) is cheapest.

B13. Put in PV = 50,000, PMT = 56,000/12 = 4666.67, FV = 0, n = 12, and compute r = 1.788. Then APY = $(1.01788)^{12} - 1$ = 23.70%

B14. Put in PV = 1 - (0.085 x 54/360) = 0.98725, FV = 1, PMT = 0, n = 54/360 = 0.15, and compute r = APY = 8.93%

B15. Put in PV = 1 - (0.11 x 180/360) = 0.945, FV = 1, n = 180/360 = 0.5, PMT = 0, and compute

r = APY = 11.98%

B16. a. Cost = Fixed cost + Float costs

Cost of check = 1.50 + (5000 × 0.08 × (5/365)) = $6.98

Cost of wire transfer = 10.00 + 0 = $10.00

b. Cost of check = 1.50 + (25,000 × 0.08 × (5/365)) = $29.28

Cost of wire transfer = 10.00 + 0 = $10.00

B17. Reduction in float = (120,000/day) × 2.5 days = $300,000

Value of float reduction = 300,000 × 0.06 = 18,000

Annual operating cost = 7500

Expected annual profit from lockbox system = 18,000 - 7500 = $10,500

B18. a. Expected profit = (20,000 × 2.5) - 30,000 = $20,000. Yes, the proposal is attractive.

b. Annual profit = (20,000 × 2.5 × 0.06) - 1000 - (0.10 × 40,000) = 3000 - 5000 = - $2000

c. The compensating balance proposal is more profitable.

C1. Alternative 1:

$$APY = \left[1 + \frac{0.12}{12}\right]^{12} - 1 = 12.68\%$$

Alternative 2: Assume amount borrowed = 0.15×(100,000 + 1500) = $116,725

Annual principal repayment = 116,725/5 = 23,345

$$0 = -100,000 + \frac{35,017.50}{(1+r)} + \frac{32,683}{(1+r)^2} + \frac{30,348.50}{(1+r)^3} + \frac{28,014}{(1+r)^4} + \frac{25,679.50}{(1+r)^5}$$

Solving the above equation for r, APY = 16.73%

C2. Alternative 2: Put in PV = 100,000, PMT = 0, FV = 128,150, n = 1, and compute APY = 28.15%

Alternative 1 is still preferred.

C3. Float is a zero sum game. EDI reduces float as well as transactions costs. The side benefiting from the float loses that benefit with EDI. Therefore, even though, in the aggregate, everyone is better off, individual firms have incentives to keep float alive when they disperse and reduce float when they collect. The result is a sort of grid lock of the status quo.

CHAPTER 20 -- ACCOUNTS RECEIVABLE AND INVENTORY MANAGEMENT

A1. **Trade credit**, is used extensively by firms, acting effectively like a loan from one firm to another. Although the loan is tied to a purchase, both sides of the transaction to lower the cost or risk of doing business. (1) Trade credit eliminates *financial intermediation* by lowering the cost of borrowing to the customer and providing the supplier a reliable short-term investment. (2) If *collateral* is repossessed, it is more valuable in the hands of the supplier who has expertise in producing, maintaining, and marketing the collateral than in the hands of other lenders such as banks. (3) *Information costs* are reduced since a supplier accumulates important information about its customers in its normal business relations and may already possess information needed to evaluate customer creditworthiness. (4) When a supplier is willing to grant credit to customers buying its products, this conveys a positive signal concerning *product quality information*. Credit provides a cheaply enforced product quality guarantee since the customer pays the trade credit on time if the product is of acceptable quality; if not, the customer ships it back and refuses to pay. (5) Trade credit helps separate the various functions of employees who authorize transactions, who physically handle products, and who handle the payments; this results in increased protection against *employee theft*. (6) When there are multiple *steps in the distribution process* and the goods pass through a number of agents, trade credit also enables the ultimate buyer to make only one payment to the original seller. (7) From the aspect of *convenience, safety, and buyer psychology*, credit eliminates the need to carry and pay with cash.

A2. The phrase 1/10, net 40 means that the invoiced amount is due within 40 days, but if the customer pays within 10 days, a discount of 1% can be taken.

A3. **Trade credit** involves zero cost to the borrower when the funds are repaid before the end of the **discount period** because the discounted price is the "actual" price of the goods purchased. The purchaser has free use of the seller's funds (equal to the discounted purchase price) from the purchase date to the payment date -- provided payment is made before the end of the discount period.

A4. The **Five C's of Credit** are five general factors that credit analysts often consider when making a credit-granting decision. (1) *Character* reflects the applicant's commitment to meet credit obligations and is best measured by the applicant's prior payment history. (2) *Capacity* is the ability to meet credit obligations with current income. (3) *Capital* represents the ability to meet credit obligations from existing assets if necessary. (4) *Collateral* can be repossessed in the case of nonpayment. (5) *Conditions* such as general economic conditions or specific industry conditions which are external to the customer's business affect the applicant's ability to repay the loan, and therefore affect the credit-granting decision.

A5. Benny Baggins's credit score is 15, so Benny meets the cutoff and qualifies for a credit card.

Telephone	4
Income	3
Employment	2
Residence	2
Credit report	4
Total score	15

A6. A typical collection process can involve the following steps. A *letter* reminding the customer that an account is overdue is sent, followed by more severe and demanding letters if payment is not received.

After the first couple of letters, a *telephone call* is made to the customer in an effort to make collection or to work out a compromise if the customer is having financial difficulties. The salesperson who made the sell or other special collectors can make a *personal visit* to the customer to request payment. The account can be turned over to a *collection agency* that specializes in collecting past due accounts. If the bill is large enough, *legal action* can be taken to obtain a judgement against the debtor.

A7. An ABC system of inventory control categorizes items from most to least important and devotes the highest degree of monitoring and control to the most important items in inventory, regardless of the volume of the item in inventory. This system conserves resources and reduces the inefficiencies associated with carefully controlling less critical items.

A8. A materials requirements planning (MRP) system is a computer based software system that plans backwards from the production schedule to make purchases and manage inventory so that production proceeds smoothly and without interruptions due to inventory stockouts..

A9. **Safety stocks** are buffer inventory held to minimize the risk of a stockout. A firm would normally find it advantageous to maintain safety stocks in its inventory because there is uncertainty regarding future demand and there are costs associated with having insufficient inventory to meet that demand.

A10. The success of a JIT system depends on: (1) a high degree of coordinated, integrated inventory *planning*; (2) good *supplier relations*; (3) controlling and reducing *setup costs*; (4) analyzing other *cost factors* such as supplier monitoring and control costs , inventory holding costs, and inventory setup costs; and (5) the *impact on credit terms* due to the elimination of accounts payable and accounts receivable through the use of EDI.

B1. **Trade credit** is used as a device to reduce financial contracting costs created by market imperfections, such as imperfect information and search difficulties, allowing both sides of the transaction to lower the cost or risk of doing business. Rather than pay higher interest rates to financial intermediaries who stand between the borrower and the lender, firms use trade credit. The interest rate of a trade credit loan is lower than customers' alternative borrowing rates, and the customer makes a reliable short-term investment for the supplier.

B2. By Equation (20.1): $\text{NPV} = \dfrac{pR}{(1+r)^t} \quad C = \dfrac{0.95 \times 200{,}000}{(1.18)^{0.25}} - (0.7 \times 200{,}000) = \$42{,}298$

B3.

$$\text{Interest rate} = \frac{12}{365} = 0.032877\%$$

$$\text{PV of expected payment} = \frac{(0.50 \times 4000 \times 0.98)}{(1.00032877)^{10}} + \frac{(0.4 \times 4000)}{(1.00032877)^{30}} = \$3537.87$$

$$\text{NPV} = \frac{p \times R}{(1+r)^t} - C = \text{PV of Expected Payment} - C = 3537.87 - 3400 = \$137.87$$

B4. By Equation (20.2)

Refrigerator: $$p* = \frac{C \times (1 + r)^t}{R} = \frac{1000 \times (1.12)^{0.25}}{1200} = 0.86$$

Jewelry: $$p* = \frac{C \times (1 + r)^t}{R} = \frac{150 \times (1.14)^1}{300} = 0.57$$

Stereo: $$p* = \frac{C \times (1 + r)^t}{R} = \frac{550 \times (1.13)^{0.5}}{800} = 0.73$$

B5.

Age	Accounts Receivable	Percent
0-30 days	$10,200	53.13%
30-60 days	6,000	31.25
60-90 days	3,000	15.62
Total	$19,200	100.00%

B6. Average age = (0.5333 × 15) + (0.3333 × 45) + (0.1333 × 75) = 33.00 days

B7. a. Sales outstanding at the end of the month of sale = 80%

Sales outstanding in the first month after sale = 30%

Sales outstanding in the second month after sale = 5%

Sales outstanding in the third month after sale = 0%

b.

Month	Accounts Receivable
October	$ 400
November	550
December	575
January	575
February	735
March	955

c.

Age	Accounts Receivable	Percent
0-30 days	$ 720	75.39%
30-60 days	210	21.99
60-90 days	25	2.62
Total	$ 955	100.00%

B8. a. Expected inflow = 10,000,000 × (1 - 0.01) = $9,900,000

Cost of sales = 10,000,000 × 0.85 = 8,500,000

NPV(current policy) = PV(expected inflows) - Cost of sales

$$= \frac{9{,}900{,}000}{(1.15)^{0.1}} - 8{,}500{,}000 = \$1{,}262{,}598$$

b. Expected inflow = 11,000,000 × (1 - 0.02) = $10,780,000

Cost of sales = 11,000,000 × 0.85 = 9,350,000

NPV(new policy) = PV(expected inflow) - Cost of sales

$$= \frac{10{,}780{,}000}{(1.15)^{0.2}} - 9{,}350{,}000 = \$1{,}132{,}846$$

c. The current policy is more profitable since it has a higher NPV.

B9. By Equation (20.3): Total Cost = Ordering cost + Carrying cost $= F \times \frac{S}{Q} + C \times \frac{Q}{2}$

Average inventory = Q/2

Orders per year = Annual usage/order size = Annual usage/Q

Order size (units)	200	400	600	800	1000
Average inventory	100	200	300	400	500
Orders per year	60	30	20	15	12
Annual cost ordering	3600	1800	1200	900	720
Annual carrying cost	400	800	1200	1600	2000
Total annual cost	4000	2600	2400	2500	2720

B10. a. By Equation (20.3), annual cost of ordering 6 rolls at a time:

$$\text{Total Cost} = F \times \frac{S}{Q} + C \times \frac{Q}{2} = [500 \times \frac{6 \times 52}{6}] + [1000 \times \frac{6}{2}] = \$26{,}300$$

b. By Equation (20.4):

$$Q = \text{EOQ} = \left[\frac{2FS}{C}\right]^{\frac{1}{2}} = \left[\frac{2 \times 500 \times 312}{1000}\right]^{\frac{1}{2}} = 17.66$$

c. Annual cost of ordering 18 rolls at a time:

$$\text{Total Cost} = [500 \times \frac{6 \times 52}{18}] + [1000 \times \frac{6}{2}] = \$8966$$

Annual cost of ordering 17 rolls at a time:

$$\text{Total Cost} = [500 \times \frac{6 \times 52}{17}] + [1000 \times \frac{6}{2}] = \$9476$$

The optimal order size is 18 rolls since it has the lower total annual cost.

B11. a. By Equation (20.4):

$$Q = \text{EOQ} = \sqrt{\frac{2 \times F \times S}{C}} = \sqrt{\frac{2 \times 50 \times 10{,}000}{4}} = 500$$

Annual cost of ordering 500 at a time by Equation (20.3):

$$\text{Total Cost} = F \times \frac{S}{Q} + C \times \frac{Q}{2} = [50 \times \frac{10{,}000}{500}] + [4 \times \frac{500}{2}] = \$2000$$

b. Order the EOQ of 500 and receive the discount for ordering more than 250 at a time.

$$\text{Total Cost} = F \times \frac{S}{Q} + C \times \frac{Q}{2} - d \times S = [50 \times \frac{10{,}000}{500}] + [4 \times \frac{500}{2}] - 0.10 \times 10{,}000 = \$1000$$

c. Order the 1000 minimum required to get the discount:

$$\text{Total Cost} = F \times \frac{S}{Q} + C \times \frac{Q}{2} - d \times S = [50 \times \frac{10{,}000}{500}] + [4 \times \frac{1000}{2}] - 0.10 \times 10{,}000 = \$500$$

Recommend ordering 1000 switches at a time because this order size has a lower annual cost.

B12. a. Expected lead time demand = lead time × average usage = 4 × 4 = 16 kits

b. Reorder point = Expected lead time demand + Safety Stock = 16 + 10 = 26 kits

B13. Total Cost = Ordering Cost + Carrying Cost + Stockout Cost

Safety Stock	Total Cost
0	$ 32,000
5	$ 29,500
10	$ 29,000
15	$ 29,500
20	$ 31,000

Total annual costs are minimized using a safety stock of 10 and a reorder point of 30.

B14. a. By Equation (20.4):

$$EOQ = \sqrt{\frac{2FS}{C}} = \sqrt{\frac{2 \times 500 \times 1000 \times 12}{12}} = 1000$$

b. Carrying costs of stockout = safety stock × 12

Each stockout costs (0.8 × 30) + 10 = $34.

Safety Stock	Stockout Cost	Carrying Cost	Total Cost of Stockout
100	75 × 34 = 2550	1200	$3750
125	50 × 34 = 1700	1500	$3200
150	35 × 34 = 1190	1800	$2990
175	30 × 34 = 1020	2100	$3120
200	30 × 34 = 1020	2400	$3420

The minimum total cost of stockout is $2990, so the optimal level of safety stock is 150 tires.

C1. The firm's inventory management problem could be modeled as a capital investment problem by choosing a particular investment horizon, computing the initial outlay for the inventory investment, specifying a particular inventory policy, estimating the incremental sales revenue that would result under that policy, and estimating the amounts and timing of costs associated with the proposed policy (ordering costs, carrying costs, expected stockout costs, etc.). The net present value of the proposed alternative would be computed by discounting the incremental free cash flow at the opportunity cost of inventory investment and subtracting the initial outlay. The set of possible inventory management policies is mutually exclusive. The firm should choose the inventory management policy that has the largest NPV.

Similarly, the firm's accounts receivables management problem could be modeled as a capital investment problem by computing the NPV for each alternative receivables policy. For example, in analyzing the decision as to whether to offer a trade discount, the firm would estimate incremental cash flows associated with the discount policy (incremental sales, cost of the discount, cost of additional clerical staff, etc.), analyze the impact of the discount on the receivables collection period, and then discount the cash flows at the opportunity cost of receivables investment to find the NPV of implementing the discount. The set of receivables management policies for a particular item and a particular class of customers is mutually exclusive. The firm should therefore choose the receivables management policy that has the largest NPV.

C2. a. With an order quantity of Q units and sales of S units per year, the number of orders placed per year is S/Q. With a fixed cost per order of F, the annual reordering cost is $F \times S/Q$.

b. With an order quantity of Q units and constant rate of sale, the average inventory will be Q/2 units. With annual holding costs of C per unit, the annual inventory holding cost is C × Q/2.

c. The aggregate annual inventory cost (AC) is the sum of the holding cost and the reordering cost:

$$AC = (C \times Q/2) + (F \times S/Q)$$

d. To minimize aggregate cost, take the derivative with respect to Q, set it equal to zero, and solve for Q:

$$\frac{\partial AC}{\partial Q} = \frac{C}{2} - \frac{F \times S}{Q^2} = 0$$

So that, $C/2 = F \times S/Q^2$; and $Q^2 = 2 \times F \times S/C$

$$Q = \sqrt{\frac{2 \times F \times S}{C}} = \text{EOQ}$$

C3. Month:	Jan	Feb	Mar	April	May	June	July
Sales:	1500	1800	2000	2200	1400	1700	1300
Panel A							
Collections from:							
Current month	135[a]	160[a]	175[a]	198	126	153	117
Previous month		680[a]	800[a]	900	990	630	765
Two months previous			525[a]	648	720	792	504
Three month previous				160	192	205	220
Total collections				1906	2028	1780	1606
Panel B							
Collection fractions from:							
Current month	0.090[a]	0.089[a]	0.088[a]	0.090	0.090	0.090	0.090
Previous month		0.453[a]	0.444[a]	0.450	0.450	0.450	0.450
Two months previous			0.350[a]	0.360	0.360	0.360	0.360
Three month previous				0.107[b]	0.107[b]	0.102[b]	0.100
Panel C							
Accounts receivable from:							
Current month	1365[a]	1640[a]	1825[a]	2002	1274	1547	1183
Previous month		685[a]	840[a]	920	1012	644	782
Two months previous			160[a]	180	200	220	140
Three month previous				0	0	0	0
Total accounts receivable				3102	2486	2411	2105
Panel D							
Receivables balance fractions from:							
Current month	0.910[a]	0.911[a]	0.9125[a]	0.910	0.910	0.910	0.910
Previous month		0.457[a]	0.467[a]	0.460	0.460	0.460	0.460
Two months previous			0.107[a]	0.100	0.100	0.100	0.100
Three month previous				0.000	0.000	0.000	0.000

[a]This information is taken from Table 20-3 on page 643 of the text.
[b]This amount is put in to force the collection fractions to sum to 1.0.

CHAPTER 21 - TREASURY MANAGEMENT

A1. The controller is responsible for financial accounting, managerial/cost accounting, internal auditing, tax accounting, accounts payable and information systems.

A2. The treasurer is responsible for cash and marketable securities management, capital budgeting, financial planning, credit analysis, investor relations, risk and insurance management, and pension fund management.

A3. Under a defined benefit pension plan, the promised retirement payments are based on the employee's years of service to the firm and final salary.

A4. Under a defined contribution pension plan, every employee has an account to which the employer makes regular contributions. In some cases, employees can also contribute to the account.

A5. Using Equation (21.1): Annual pension payment = Years of service×Percent factor×Final salary
Annual pension payment = 28×1.9%×90,000 = \$47,880.

A6. The major provisions of ERISA (Employee Retirement Income Security Act of 1974) include: (1) minimum funding standards--the firm is required to make minimum pension contributions based on projected obligations; (2) fiduciary standards--the law establishes fiduciary standards for loyalty, prudence, and diversification that pension fund trustees must follow; (3) minimum vesting standards--ERISA set minimum requirements for the conferring of pension rights on employees; and (4) the creation of the Pension Benefit Guarantee Corporation (PBGC)--a government agency that insures vested benefits of defined benefit plans within certain limits.

A7. The tax system favors overfunding because contributions are tax deductible and pension fund earnings accumulate tax free. Earnings are taxed when they are distributed to beneficiaries.

A8. Yes. The increase in spending is beneficial. The firm would save \$300M×0.04% = 100,000 annually on debt and \$250M×0.12% = 360,000 annually on equity. Total amount saved equals 460,000. The firm has a net benefit of 460,000 - 250,000 = \$210,000 from the IR program.

A9. The four forms of financial communication are computer and telecommunications technology, written communications with stockholders, the annual meeting, and communications with top management.

B1. a. Using Equation (21.1):

Annual pension payment = 40×3.1%×150,000 = \$186,000 per year (15,500 per month)

b. He is expected to live 25 years (300 months) and the required return is 6% (0.5% per month). The present value is \$2,405,706 [Put in CF = 15,500; r = 0.5%; n = 300; and FV = 0. Then, compute PV = 2,405,706.]:

$$PV = 15{,}150 \times \left[\frac{(1.005)^{300} - 1}{0.005 \times (1.005)^{300}} \right] = \$2{,}405{,}706$$

B2. a. Using Equation (21.1):

Annual pension payment = $20 \times 1.3\% \times 50{,}000$ = $13,000 per year.

At retirement, she gets 13,000 from the pension and 15,000 from Social Security, for a total of $28,000 per year. The proportion is 56% of her current income (=28,000/50,000).

b. Social Security is indexed to inflation and becomes $3 \times 15{,}000 = 45{,}000$. The pension remains at 13,000, so total income is 58,000. Keeping the pre-retirement standard of living requires $3 \times 50{,}000 = 150{,}000$. The proportion becomes 39% (= 58,000/150,000).

B3. a. Using Equation (21.1):

Annual pension payment = $10 \times 2.0\% \times [50{,}000 \times (1.05)^5]$ = $12,762.82 per year.

This assumes the vested employee quits today (current years of service) and that inflation in wages is built into the funding requirements (grow salary).

b. Life expectancy is 15 years; the required return is 7%. The present value is $116,242.67 [Put in CF = 12,762.82; r = 7%; n = 15; and FV = 0. Compute PV = 116,242.67.]:

$$PV = 12{,}762.82 \times \left[\frac{(1.07)^{15} - 1}{0.07 \times (1.07)^{15}}\right] = \$116{,}242.67$$

This is the value of the pension when Gordon retires (in 5 years), assuming he quits today and inflation in wages is built into the funding. Thus, to get the present value of the defined benefit plan, discount the 116,242.67 back 5 years. The present value is $82,879 [Put in FV = 116,242.67; r = 7%; n = 5; and CF = 0. Compute PV = 82,879]:

$$PV = \frac{116{,}242.67}{(1.07)^5} = \$82{,}879$$

So the firm has a fund surplus of $17,121.

B4. With a defined benefit plan, the firm promises to pay a fixed amount for the employee's life, which is a substantial financial obligation. The firm also bears the investment risk of the plan. While these are disadvantages to the firm, they are advantages to the employee. A disadvantage to the employee is that the defined benefit plan does not index benefits for inflation. With a defined contribution plan, the firm has no obligation beyond making regular contributions, so the employee is not guaranteed a fixed amount upon retirement. The employee also bears the investment risk of the plan. While these are advantages to the firm, they are disadvantages to the employee. The advantages are that the employee can usually make additional contributions and can select the risk level by deciding how to allocate the funds across the investment alternatives.

B5. a. Using Equation (21.1):

Wanda: Annual pension payment = $42 \times 2.1\% \times [30{,}000 \times (1.07)^{40}]$ = \$396,224 per year.

Charles: Annual pension payment = $30 \times 2.1\% \times [80{,}000 \times (1.07)^{5}]$ = \$70,689 per year.

Peggy: Annual pension payment = $30 \times 2.1\% \times [60{,}000 \times (1.07)^{20}]$ = \$146,274 per year.

This assumes that they stay with the firm until retirement.

b. Using Equation (21.1):

Wanda: Annual pension payment = $2 \times 2.1\% \times [30{,}000 \times (1.07)^{40}]$ = \$18,867.82 per year

This assumes she quits today (current years of service) and that inflation in wages is built into the funding (grow salary).

Life expectancy is 18 years; the required return is 9%. The present value is \$165,199.56 [Put in CF = 18,867.82; r = 9%; n = 18; and FV = 0. Compute PV = 165,199.56.]:

$$PV = 18{,}867.82 \times \left[\frac{(1.09)^{18} - 1}{0.09 \times (1.09)^{18}}\right] = \$165{,}199.56$$

This is the value of the pension when Wanda retires. To get the present value of the plan, discount the 165,199.56 back 40 years. The present value is \$5260 [Put in FV = 165,199.56; r = 9%; n = 40; and CF = 0. Compute PV = 5260]:

$$PV = \frac{165{,}199.56}{(1.09)^{40}} = \$5260$$

Charles: Annual pension payment = $25 \times 2.1\% \times [80{,}000 \times (1.07)^{5}]$ = \$58,907.17 per year.

Life expectancy is 12 years; the required return is 9%. The present value is \$421,818.06 [Put in CF = 58,907.17; r = 9%; n = 12; and FV = 0. Compute PV = 421,818.06.]:

$$PV = 58{,}907.17 \times \left[\frac{(1.09)^{12} - 1}{0.09 \times (1.09)^{12}}\right] = \$421{,}818.06$$

This is the value of his pension when he retires. To get the present value of the plan, discount the 421,818.06 back 5 years. The present value is \$274,153 [Put in FV = 421,818.06; r = 9%; n = 5; and CF = 0. Compute PV = 274,153.]:

$$PV = \frac{421{,}818.06}{(1.09)^{5}} = \$274{,}153$$

Peggy: Annual pension payment = $10\times 2.1\%\times[60{,}000\times(1.07)^{20}]$ = \$48,758.02 per year.

Life expectancy is 20 years; the required return is 9%. The present value is \$445,089.81 [Put in CF = 48,758.02; r = 9%; n = 20; and FV = 0. Compute PV = 445,089.81.]:

$$PV = 48{,}758.02\times\left[\frac{(1.09)^{20} - 1}{0.09\times(1.09)^{20}}\right] = \$445{,}089.81$$

This is the value of her pension when she retires. To get the present value of the plan, discount the 445,089.81 back 20 years. The present value is \$79,418 [Put in FV = 445,089.80; r = 9%; n = 20; and CF = 0. Compute PV = 79,418]:

$$PV = \frac{445{,}089.81}{(1.09)^{20}} = \$79{,}418$$

B6. Path 1:

	Starting		Ending
1	460,000	1.050	483,000
2	493,000	0.920	453,560
3	463,560	0.875	405,615
4	415,615	1.110	461,333
5	471,333	1.050	494,899
6	504,899	0.970	489,752
7	499,752	1.070	534,735
8	544,735	1.210	659,129
9	669,129	1.060	709,277
10	719,277	1.040	748,048

Path 2:

	Starting		Ending
1	460,000	1.100	506,000
2	516,000	1.175	606,300
3	616,300	1.140	702,582
4	712,582	0.900	641,324
5	651,324	1.030	670,864
6	680,864	0.940	640,012
7	650,012	0.930	604,511
8	614,511	1.090	669,817
9	679,817	1.030	700,212
10	710,212	1.120	759,437

B7. Total amount per year is $17,000. To get the FV after 25 years [Put in CF = annual installment; r = expected return; n = 25; and PV = 0. Compute FV]:

$$\text{FV after 25 years} = \text{Annual Installment} \times \left[\frac{(1 + \text{expected return})^{25} - 1}{\text{expected return}}\right]$$

Lacey:

	Annual installment	Expected return	FV after 25 years
Money market	6800	6%	373,079
Bond	7650	9%	647,962
Stock	2550	12%	340,001
Total	17,000		1,361,042

Logan:

	Annual installment	Expected return	FV after 25 years
Money market	1700	6%	93,270
Bond	4250	9%	359,979
Stock	11,050	12%	1,473,339
Total	17,000		1,926,588

There is a 6% APR on lifetime annuities, so the firm would quote 16.67 (= 1/.06) as the annuity factor. Lacey's annual income would be $1,361,042/16.67 = $81,663 (= [0.06]×1,361,042). Logan's annual income would be $1,926,588/16.67 = $115,595 (= [0.06]×1,926,588).

B8. The option to default gives high-risk firms an incentive to underfund the plan. When the firm goes bankrupt, it defaults on all of its claims (including the pension). This jeopardizes the benefit payments and shifts the responsibility for the plan to the PBGC and the beneficiaries.

B9. a. Note that the investments were made at the beginning of the year. Therefore, they have an "extra" year of investment, and their present value is $66,034:

$$FV = 5000 \times \left[\frac{(1.05)^{10} - 1}{0.05}\right] \times (1.05) = \$66{,}034$$

[Alternatively, because this is an annuity due (payments at the beginning), you can set you calculator to the BEGIN mode. Then, put in CF = 5000; r = 5%; n = 10; PV = 0, and compute FV = 66,034. If you use this method, don't forget to re-set the mode back to END.]

b. Again, pavements were made at the beginning of the year, and Barbra's payments have an "extra" year of investment. Therefore their present value is $1,490,634:

$$FV = 5000 \times \left[\frac{(1.10)^{15} - 1}{0.10}\right] \times (1.10) = \$1{,}490{,}634$$

[Alternatively, because this is an annuity due (payments at the beginning), you can set you calculator to the BEGIN mode. Then, put in CF = 5000; r = 10%; n = 35; PV = 0, and compute FV = 1,490,634. If you use this method, don't forget to re-set the mode back to END.]:

B10. For diversified stakeholders, risk-reducing activities do not affect the required return, but they can still lead to improvements in firm value. The amount of operating cash flow is connected with suppliers, customers, employees, and other stakeholders; and it is unlikely that all of them will be diversified. By reducing risk, the cash flows improve, benefiting all stakeholders. Non-diversified stakeholders benefit from both improved cash flows and a reduction in total risk.

B11. Fund Balance:

Year	Starting Value	Factor	Year End Contribution	Ending Value
1	440,000	1.120	5000	497,800
2	497,800	1.180	5000	592,404
3	592,404	0.910	7500	546,588
4	546,588	1.030	7500	570,485
5	570,485	1.240	7500	714,902
6	714,902	1.040	7500	750,998
7	750,998	1.150	10,000	873,647
8	873,647	0.950	10,000	839,965
9	839,965	1.080	10,000	917,162
10	917,162	1.050	10,000	973,020

B12. Financial distress changes managerial incentives. Managers will attempt desperate strategies in an attempt to keep the firm afloat. Managers may seek short-term profits, regardless of long-term effects. They may liquidate or sell-off promising lines of business. They may sacrifice product quality or safety. Hoping for a quick payoff, managers may invest in very high-risk projects. If the firm is unlikely to survive, managers may pass up good investment projects with long-run payoffs. Financial distress gives managers strong incentives to make bad decisions.

B13. a. The expected loss is $0.20\% \times 300{,}000 = \$600$.

b. The equity value is $300{,}000 - 215{,}000 = 85{,}000$. The expected loss to the owners is $0.20\% \times 85{,}000 = 170$.

c. No. The building is the only collateral on the loan. In the event of a fire, the creditors incur a large loss. The insurance policy's payoffs are shared by the owners and the creditors, and the policy often costs more than the expected payoff to the owners. This gives the owners an incentive to underinsure. Thus, the debt contract requires the franchise to carry insurance.

B14. The four techniques for reducing risk are: (1) insurance--the firm can buy insurance or self-insure to protect against a range of hazards; (2) financial contracting--the firm can use financial futures and forward contracts as well as options to hedge against changes in prices, interest rates, and exchange rates; (3) product choice--the firm can avoid risky markets and products; and (4) leverage--the firm can reduce the amount of financial leverage that it uses. Operating contracts can also be used to reduce risk; for example, leases can be used for short-term needs.

B15. Managers communicate with financial analysts to insure accurate earnings forecasts and recommendations. Meetings provide an opportunity for investors and analysts to get to know management and to learn more about the firm. It also helps the firm to disseminate information to current and potential investors in a global marketplace.

B16. The six areas are: (1) recognition and credibility in the business community--good reputation helps sales, public relations, and employee loyalty; (2) fair valuation--by providing information, the firm helps insure fair market prices; (3) well-informed constituency of professional investors--group of professional investors can assess new ideas and incorporate the information into their stock valuations; (4) cost of capital--it can be easier and cheaper to raise debt and equity financing; (5) compliance with securities laws--IR helps to assure that the firm's officers comply with all applicable regulations; and (6) information for non-financial corporate officers--interpreting market information helps corporate officers understand the financial impact of their decisions.

C1. Taxes and risk-shifting cause plans to be overfunded or underfunded. The tax system favors overfunding because contributions are tax deductible and pension fund earnings accumulate tax free. The option to default gives high-risk firms an incentive to underfund the plan. By underfunding the plan, the firm shifts the risk (and the responsibility) to the PBGC and the beneficiaries.

C2. Human capital is very difficult to diversify. Employees have a large stake in the firm. They acquire firm-specific skills that may be difficult to transfer and incur search costs when locating a new position. By reducing total risk, it is easier to recruit and retain good employees. Good employees are more productive and innovative. This helps to keep the firm competitive.

C3. Sooner or later, the bad news will be revealed to the market. This makes the investors suspicious about the information provided by the firm and hurts the firm's reputation with stockholders and creditors. Investors will be concerned that the more serious the problem, the less likely the information will be revealed. This reduces investor and creditor confidence in any IR report released by the firm. In the future, the firm could find it more difficult to obtain financing. As a result, the firm and its managers will release both types of information equally in order to preserve investor confidence.

C4. More information insures better earnings forecasts and fuels current and potential investor interest in the firm. In addition, the firm can disseminate information to investors more rapidly by computer than by mail. Investors then have more time to assess and act on this information.

CHAPTER 22 - FINANCIAL PLANNING

A1. Steve's a cash budget shows that Steve will owe $6000 in May:

Cash Budget, September - May

	Sept.	Oct.	Nov.	Dec.	Jan.	Feb.	March	April	May
Total cash receipts	500	500	500	500	500	500	500	500	500
Total cash disbursements	4000	1000	1000	1500	4000	1000	1500	1000	1500
Ending Cash Balance:									
Net cash flow	-3500	-500	-500	-1000	-3500	-500	-1000	-500	-1000
Beginning cash balance	6000	2500	2000	1500	500	0	0	0	0
Available balance	2500	2000	1500	500	-3000	-500	-1000	-500	-1000
Monthly borrowing					3000	500	1000	500	1000
Monthly repayment									
Ending cash balance	2500	2000	1500	500	0	0	0	0	0
Cumulative loan balance	0	0	0	0	3000	3500	4500	5000	6000

A2. Tulsa Well Supply's cash budget, July-December is:

	July	August	September	October	November	December
Ending Cash Balance:						
Net cash flow	220,000	205,000	-325,000	-625,000	100,000	360,000
Beginning cash balance	1,200,000	1,420,000	1,625,000	1,300,000	1,000,000	1,000,000
Available balance	1,420,000	1,625,000	1,300,000	675,000	1,100,000	1,360,000
Monthly borrowing				325,000		
Monthly repayment					100,000	225,000
Ending cash balance	1,420,000	1,625,000	1,300,000	1,000,000	1,000,000	1,135,000
Cumulative loan balance	0	0	0	325,000	225,000	0

A3. The benefits of financial planning include: (1) it standardizes assumptions to insure consistency in decision making; (2) a future orientation provides direction for the firm and encourages the development of new ideas; (3) it helps to insure objectivity in decision making; (4) by soliciting input, it motivates employee coordination and cooperation in achieving the firm's goals; (5) lenders may require financial plans for borrowing; (6) it provides better performance evaluation, so managers are judged by their decisions rather than luck; and (7) it allows the firm to make contingency plans to prepare for unlikely outcomes.

A4. All asset accounts vary directly with sales, so additional assets = cash + accounts receivable + inventory + net fixed assets = 3% + 20% + 15% + 25% = 63% of the increase in sales. Similarly, the increase in short-term liabilities (accounts payable and accruals) equals 15% of the sales increase. Net income = $0.06\times15{,}000{,}000 = 900{,}000$ and $D = 0.25\times900{,}000 = 225{,}000$. Using Equation (22.1):

$$AFN = (A/S)\times g\times S_0 - (L/S)\times g\times S_0 - [M\times(1 + g)\times S_0 - D]$$

$$= (0.63)\times 0.25\times 12{,}000{,}000 - (0.15)\times 0.25\times 12{,}000{,}000 - [0.06 \times(1.25)\times 12{,}000{,}000 - 225{,}000]$$

$$= 1{,}890{,}000 - 450{,}000 - 675{,}000 = \$765{,}000.$$

All of the additional financing is from the short-term loan.

Pro forma balance sheet for Amalgamated Meat Loaf:

Assets:	Current	Projected	Change
Cash	360,000	450,000	90,000
Accounts receivable	2,400,000	3,000,000	600,000
Inventory	1,800,000	2,250,000	450,000
Total current assets	4,560,000	5,700,000	1,140,000
Net fixed assets	3,000,000	3,750,000	750,000
Total assets	7,560,000	9,450,000	1,890,000

Liabilities and Equity:	Current	Projected	Change
Accounts Payable	1,200,000	1,500,000	300,000
Accruals	600,000	750,000	150,000
Short-term Loan	800,000	1,565,000	765,000
Total current liabilities	2,600,000	3,815,000	1,215,000
Long-term debt	1,000,000	1,000,000	0
Stockholders' equity	3,960,000	4,635,000	675,000
Total liabilities & equity	7,560,000	9,450,000	1,890,000

A5. Using Equation (22.1):

AFN = required increase in assets - increase in liabilities - increase in retained earnings

= 750,000 - 175,000 - [225,000 - 0] = $350,000

A6. Using Equation (22.1):

AFN = required increase in assets - increase in liabilities - increase in retained earnings

= 100 - 10 - [70 - D] = 50

The largest dividend is $30.

B1. Cash budget for South Carolina Sportsplex, January - June:

	Jan.	Feb.	Mar.	Apr.	May	June
Total cash receipts	100	120	150	320	350	240
Total cash disbursements	150	180	200	290	240	180
Ending Cash Balance:						
Net cash flow	-50	-60	-50	30	110	60
Beginning cash balance	150	100	100	100	100	130
Available balance	100	40	50	130	210	190
Monthly borrowing		60	50			
Monthly repayment				30	80	
Ending cash balance	100	100	100	100	130	190
Cumulative loan balance	0	60	110	80	0	0

B2. Pro forma balance sheet for Columbus Distributor:

Assets:	Current	Projected	Change
Cash	500,000	700,000	200,000
Accounts receivable	3,500,000	4,900,000	1,400,000
Inventory	3,000,000	4,200,000	1,200,000
Net fixed assets	3,000,000	3,500,000	500,000
Total assets	10,000,000	13,300,000	3,300,000

Liabilities and Equity:	Current	Projected	Change
Accounts Payable	1,000,000	1,400,000	400,000
Bank loan	1,500,000	2,000,000	500,000
Long-term bond	2,000,000	1,800,000	-200,000
Stockholders' equity	5,500,000	6,660,000	1,160,000
Total liabilities & equity	10,000,000	11,860,000	1,860,000

Since sales grow by 40%, accounts that increase proportionately with sales also rise by 40%. For example, projected cash = current cash×(1+g) = 500,000×1.40 = 700,000. Net income = 0.06×21,000,000 = 1,260,000 and retained earnings increases by 1,260,000 - 100,000 = 1,160,000. The balance sheet shows that the firm needs 3,300,000 - 1,860,000 = $1,440,000 in additional external funds.

B3. a. A/S = 1.25 and L/S = 0.15. Sales will increase by 10% to $55,000,000. Net income = 0.08×55,000,000 = 4,400,000 and D = 0.25×4,400,000 = 1,000,000. Using Equation (22.1):

$$AFN = (A/S)\times g\times S_0 - (L/S)\times g\times S_0 - [M\times(1+g)\times S_0 - D]$$

$$= 1.25\times0.10\times50M - 0.15\times0.10\times50M - [0.08\times1.10\times50M - 1.1M]$$

$$= 6{,}250{,}000 - 750{,}000 - 3{,}300{,}000 = \$2{,}200{,}000$$

b. d = 0.25. Using Equation (22.2):

$$g = \frac{M\times(1-d)}{[(A/S) - (L/S) - M\times(1 - d)]} = \frac{0.08\times(1-0.25)}{[(1.25) - (0.15) - 0.08\times(1-0.25)]} = \frac{0.06}{1.04}$$

g = 5.77%

c. B/E = 0.60. Using the adjusted equation for AFN:

$$\text{AFN} = (A/S)\times g\times S_0 - (L/S)\times g\times S_0 - M\times S_0\times(1+g)\times(1-d)\times(1+B/E)$$

= 1.25×0.10×50M - 0.15×0.10×50M - 0.08 ×50M×1.10×(1-0.25)×(1.60)

= 6,250,000 - 750,000 - 5,280,000 = $220,000

d. Using Equation (22.3):

$$g = \frac{M\times(1-d)\times(1+B/E)}{[(A/S) - (L/S) - M\times(1 - d)(1+B/E)]}$$

$$g = \frac{0.08\times(1-0.25)\times(1.60)}{[(1.25) - (0.15) - 0.08\times(1-0.25)\times(1.60)]} = \frac{0.096}{1.004} = 9.56\%$$

B4. Cash budget for Merrimack Resorts, April - September:

	April	May	June	July	Aug.	Sept.
Total cash receipts	100	100	175	250	300	250
Total cash disbursements	200	200	250	250	175	125
Ending Cash Balance:						
Net cash flow	-100	-100	-75	0	125	125
Beginning cash balance	150	100	100	100	100	100
Available balance	50	0	25	100	225	225
Monthly borrowing	50	100	75			
Monthly repayment					125	100
Ending cash balance	100	100	100	100	100	125
Cumulative loan balance	50	150	225	225	100	0

B5. Dark Adventures cash budget:

Month	Oct.	Nov.	Dec.	Jan.	Feb.	Mar.
Sales	600,000	700,000	800,000	800,000	700,000	600,000
Cash Receipts:						
Collections (current) 25%	150,000	175,000	200,000	200,000	175,000	150,000
(previous month) 50%	250,000	300,000	350,000	400,000	400,000	350,000
(2nd month previous) 25%	100,000	125,000	150,000	175,000	200,000	200,000
Total cash receipts	500,000	600,000	700,000	775,000	775,000	700,000
Cash Disbursements:						
Purchases	350,000	400,000	400,000	350,000	300,000	250,000
Wages and salaries	144,000	168,000	192,000	192,000	168,000	144,000
Rent	4000	4000	4000	4000	4000	4000
Cash operating expenses	8000	8000	20,000	20,000	8000	8000
Tax installments	10,000			10,000		
Capital expenditure	100,000					
Mortgage Payment	5000	5000	5000	5000	5000	5000
Total cash disbursements	621,000	585,000	621,000	581,000	485,000	411,000
Ending Cash Balance:						
Net cash flow	-121,000	15,000	79,000	194,000	290,000	289,000
Beginning cash balance	125,000	60,000	60,000	98,000	292,000	582,000
Available balance	4000	75,000	139,000	292,000	582,000	871,000
Monthly borrowing	56,000					
Monthly repayment		15,000	41,000			
Ending cash balance	60,000	60,000	98,000	292,000	582,000	871,000
Cumulative loan balance	56,000	41,000	0	0	0	0

B6. a. Pro forma balance sheet ($thousands) for Kennesaw Leisure Products (10% Sales increase):

Assets:	Current	Projected	Change
Cash	50	55	5
Accounts receivable	220	242	22
Inventory	300	330	30
Net fixed assets	210	210	—
Total assets	780	837	57

Liabilities and Equity:	Current	Projected	Change
Accounts Payable	70	77	7
Bank loan	180	180	
Long-term bond	200	190	-10
Stockholders' equity	330	396	66
Total liabilities & equity	780	843	63

Since sales grow by 10%, accounts that increase proportionately also rise by 10%. For example, projected cash = current cash$\times(1 + g) = 50\times1.10 = 55$. Sales will increase by 10% to $660,000 next year. Net income = $0.10\times660{,}000 = 66{,}000$. The balance sheet shows that the firm has excess funds of 63 - 57 = $6 (thousand) available.

b. Pro forma balance sheet ($ thousands) for Kennesaw Leisure Products (20% sales increase):

Assets:	Current	Projected	Change
Cash	50	60	10
Accounts receivable	220	264	44
Inventory	300	360	60
Net fixed assets	210	210	
Total assets	780	894	114

Liabilities and Equity:	Current	Projected	Change
Accounts Payable	70	84	14
Bank loan	180	180	
Long-term bond	200	190	-10
Stockholders' equity	330	396	66
Total liabilities & equity	780	850	70

Now the proportionate accounts will rise by 20%. The balance sheet shows that the firm needs 114 - 70 = $44 (thousand) in additional external funds.

B7. A/S = 1.50 and L/S = 0.15. Sales will increase by 20% to $1,200,000. Net income = 0.20×1,200,000 = 240,000 and D = 240,000/3 = 80,000 Using Equation (22.1):

$$AFN = (A/S)\times g\times S_0 - (L/S)\times g\times S_0 - [M\times(1+g)\times S_0 - D]$$

$$= 1.50\times0.20\times1{,}000{,}000 - 0.15\times0.20\times1{,}000{,}000 - [0.20\times1.20\times1{,}000{,}000 - 80{,}000]$$

$$= 300{,}000 - 30{,}000 - 160{,}000 = \$110{,}000$$

B8. a. A/S = 1.00 and L/S = 0.20. Using Equation (22.2):

$$g = \frac{M\times(1-d)}{[(A/S) - (L/S) - M\times(1-d)]} = \frac{0.09\times(1-0.5)}{[(1.00) - (0.20) - 0.09\times(1-0.5)]} = \frac{0.045}{0.755} = 5.96\%$$

b. B/E = 0.75. Using Equation (22.3):

$$g = \frac{M\times(1-d)\times(1+B/E)}{[(A/S) - (L/S) - M\times(1-d)\times(1+B/E)]}$$

$$g = \frac{0.09\times(1-0.5)\times(1.75)}{[(1.00) - (0.20) - 0.09\times(1-0.5)\times(1.75)]} = \frac{0.07875}{0.72125} = 10.92\%$$

B9. a. The two sources of internal funds are: 1) short-term, naturally occurring liabilities (such as accounts payable and accruals); and 2) new retained earnings. The firm can also decrease dividends in order to keep more retained earnings for reinvestment.

b. The additional source of funds is new external debt.

B10. An increase in the variables would cause the internal growth rate to:

<u>Increase</u>	net profit margin
<u>Decrease</u>	dividend payout ratio
<u>Increase</u>	asset requirement as a fraction of sales (A/S)
<u>Decrease</u>	spontaneous short-term financing as a fraction of sales (L/S)

C1.

M. T. BOX COMPANY 000'S OMITTED THROUGHTOUT

CASH BUDGET FOR JANUARY THROUGH JUNE

	OCT	NOV	DEC	JAN	FEB	MAR	APR	MAY	JUN (CASH BUDGET)	JUL	AUG
SALES	60.00	90.00	150.00	250.00	235.00	190.00	160.00	110.00	90.00	135.00	260.00
A/R COLLECTIONS				127.40	175.10	224.25	212.25	171.65	139.60		
INTEREST INCOME				0.68	0.11	0.09	0.58	0.21	0.72		
CASH INFLOW				128.08	175.21	224.34	212.83	171.86	140.32		
PAYABLES PAID				160.00	145.25	122.30	93.70	71.80	78.55		
SALARIES & WAGES PAID				50.00	47.90	41.60	37.40	30.40	27.60		
TAXES PAID				5.96	4.84	2.02	0.32	-3.47	-4.52		
INTEREST PAID				0.00	0.00	15.00	0.00	0.00	15.00		
FIXED ASSET PURCHASES							150.00				
DIVIDENDS PAID				2.00	2.00	2.00	2.00	2.00	2.00		
STARTING CASH				170.00	155.07	133.77	109.59	88.61	91.85		
ENDING CASH				155.07	133.77	109.59	88.61	91.85	113.96		
S-T INVESTMENT CHANGE (cash budget)				-74.96	-3.47	65.60	-49.62	67.88	-0.42		
S-T INVESTMENT CHANGE (balance sheet)				-74.96	-3.47	65.60	-49.62	67.88	-0.42		

MONTHLY PRO FORMA INCOME STATEMENTS

	OCT	NOV	DEC	JAN	FEB	MAR	APR	MAY	JUN (INCOME STATEMENT)	JUL	AUG
SALES	60.00	90.00	150.00	250.00	235.00	190.00	160.00	110.00	90.00	135.00	260.00
COST OF RAW MATERIALS				170.00	159.80	129.20	108.80	74.80	61.20		
SALARIES & WAGES				50.00	47.90	41.60	37.40	30.40	27.60		
DEPRECIATION				6.15	6.09	6.03	5.97	7.41	7.33		
INTEREST EXPENSE				7.50	7.50	7.50	7.50	7.50	7.50		
INTEREST INCOME				0.68	0.11	0.09	0.58	0.21	0.72		
TAXABLE INCOME				17.03	13.82	5.76	0.91	-9.90	-12.92		
TAXES				5.96	4.84	2.02	0.32	-3.47	-4.52		
NET INCOME				11.07	8.99	3.74	0.59	-6.44	-8.40		
DIVIDENDS				2.00	2.00	2.00	2.00	2.00	2.00		
CHANGE IN RETAINED EARNINGS				9.07	6.99	1.74	-1.41	-8.44	-10.40		

CURRENT BALANCE SHEET AND MONTHLY PRO FORMA BALANCE SHEETS

	OCT	NOV	DEC	JAN	FEB	MAR	APR	MAY	JUN (BALANCE SHEET)	JUL	AUG
CASH			170.00	155.07	133.77	109.59	88.61	91.85	113.96		
SHORT-TERM INVESTMENT (BORROWING)			90.00	15.04	11.58	77.18	27.56	95.44	95.02		
ACCOUNTS RECEIVABLE			195.00	317.60	377.50	343.25	291.00	229.35	179.75		
INVENTORY			157.00	132.25	94.75	59.25	22.25	26.00	96.75		
NET FIXED ASSETS			615.00	608.85	602.76	596.73	740.77	733.36	726.03		
TOTAL ASSETS			1227.00	1228.82	1220.35	1186.00	1170.19	1176.00	1211.51		
ACCOUNTS PAYABLE			160.00	145.25	122.30	93.70	71.80	78.55	131.95		
ACCRUED INTEREST			0.00	7.50	15.00	7.50	15.00	22.50	15.00		
LONG-TERM DEBT			600.00	600.00	600.00	600.00	600.00	600.00	600.00		
COMMON STOCK			120.00	120.00	120.00	120.00	120.00	120.00	120.00		
RETAINED EARNINGS			347.00	356.07	363.05	364.80	363.39	354.95	344.56		
LIABS. + OWNERS' EQUITY			1227.00	1228.82	1220.35	1186.00	1170.19	1176.00	1211.51		

C2.

DUNN MANUFACTURING 000'S OMITTED THROUGHTOUT

CASH BUDGET FOR APRIL THROUGH DECEMBER

										CASH BUDGET		
	JAN	FEB	MAR	APR	MAY	JUN	JUL	AUG	SEP	OCT	NOV	DEC
SALES	320.00	345.00	365.00	410.00	430.00	350.00	325.00	300.00	220.00	265.00	290.00	375.00
A/R COLLECTIONS				348.60	371.60	383.70	388.80	344.90	299.50	283.50	248.50	276.10
PAYABLES PAID				177.00	244.50	229.50	205.00	166.75	173.00	175.50	218.75	221.00
SALARIES & WAGES PAID				68.00	70.00	62.00	59.50	57.00	49.00	53.50	56.00	64.50
TAXES PAID						90.00						
L-T INTEREST PAID				0.00	0.00	19.50	0.00	0.00	19.50	0.00	0.00	19.50
S-T INTEREST PAID				2.41	1.82	1.08	1.11	2.50	1.24	0.73	0.57	1.00
DIVIDENDS PAID				5.00	5.00	5.00	5.00	5.00	5.00	5.00	5.00	5.00
FIXED ASSET PURCHASES							275.00					
STARTING CASH				210.00	252.38	235.53	215.40	184.96	184.05	189.08	223.26	230.90
ENDING CASH				252.38	235.53	215.40	184.96	184.05	189.08	223.26	230.90	242.84
CHANGE IN NOTES P (cash budget)				-53.82	-67.13	3.25	126.38	-114.56	-46.73	-14.58	39.46	46.84
CHANGE IN NOTES P (balance sheet)				-53.82	-67.13	3.25	126.38	-114.56	-46.73	-14.58	39.46	46.84

MONTHLY PRO FORMA INCOME STATEMENTS

										INCOME STATEMENT		
	JAN	FEB	MAR	APR	MAY	JUN	JUL	AUG	SEP	OCT	NOV	DEC
SALES	320.00	345.00	365.00	410.00	430.00	350.00	325.00	300.00	220.00	265.00	290.00	375.00
COST OF RAW MATERIALS				266.50	279.50	227.50	211.25	195.00	143.00	172.25	188.50	243.75
SALARIES & WAGES				68.00	70.00	62.00	59.50	57.00	49.00	53.50	56.00	64.50
DEPRECIATION				6.07	6.01	5.96	5.90	8.33	8.25	8.18	8.10	8.03
L-T INTEREST EXPENSE				6.50	6.50	6.50	6.50	6.50	6.50	6.50	6.50	6.50
S-T INTEREST EXPENSE				2.41	1.82	1.08	1.11	2.50	1.24	0.73	0.57	1.00
TAXABLE INCOME				60.53	66.17	46.96	40.73	30.67	12.01	23.84	30.33	51.22
TAXES				20.58	22.50	15.97	13.85	10.43	4.08	8.11	10.31	17.41
NET INCOME				39.95	43.67	31.00	26.88	20.24	7.92	15.74	20.02	33.80
DIVIDENDS				5.00	5.00	5.00	5.00	5.00	5.00	5.00	5.00	5.00
CHANGE IN RETAINED EARNINGS				34.95	38.67	26.00	21.88	15.24	2.92	10.74	15.02	28.80

CURRENT BALANCE SHEET AND MONTHLY PRO FORMA BALANCE SHEETS

										BALANCE SHEET		
	JAN	FEB	MAR	APR	MAY	JUN	JUL	AUG	SEP	OCT	NOV	DEC
CASH			210.00	252.38	235.53	215.40	184.96	184.05	189.08	223.26	230.90	242.84
ACCOUNTS RECEIVABLE			256.00	317.40	375.80	342.10	278.30	233.40	153.90	135.40	176.90	275.80
INVENTORY			437.00	415.00	365.00	342.50	298.00	276.00	308.50	355.00	387.50	383.00
PREPAID TAXES			43.00	22.42	-0.08	73.96	60.11	49.68	45.60	37.49	27.18	9.77
OTHER ASSETS			15.00	15.00	15.00	15.00	15.00	15.00	15.00	15.00	15.00	15.00
NET FIXED ASSETS			674.00	667.93	661.92	655.97	925.06	916.74	908.49	900.31	892.21	884.18
TOTAL ASSETS			1635.00	1690.13	1653.17	1644.92	1761.43	1674.86	1620.56	1666.46	1729.69	1810.58
ACCOUNTS PAYABLE			177.00	244.50	229.50	205.00	166.75	173.00	175.50	218.75	221.00	239.25
NOTES PAYABLE (INVESTED)			219.00	165.18	98.05	101.30	227.68	113.12	66.39	51.81	91.27	138.11
ACCRUED INTEREST			0.00	6.50	13.00	0.00	6.50	13.00	0.00	6.50	13.00	0.00
LONG-TERM DEBT			650.00	650.00	650.00	650.00	650.00	650.00	650.00	650.00	650.00	650.00
COMMON STOCK			150.00	150.00	150.00	150.00	150.00	150.00	150.00	150.00	150.00	150.00
RETAINED EARNINGS			439.00	473.95	512.62	538.62	560.50	575.74	578.66	589.40	604.42	633.22
LIABS. + OWNERS' EQUITY			1635.00	1690.13	1653.17	1644.92	1761.43	1674.86	1620.56	1666.46	1729.69	1810.58

January's SALARIES & WAGES PAID are projected to be 62.00, but are not shown.

CHAPTER 23 -- EQUITY AND THE INVESTMENT BANKING PROCESS

A1. Common stock is a perpetual security which is not redeemable. Preferred stock typically is redeemable. Common stock dividends are not stated obligations, and common stockholders receive dividends only when they are declared by the board of directors. Preferred dividends are paid at a stated rate. Common stock carries with it the right to vote for the board of directors. Preferred stock does not have similar voting rights.

A2. Public financing involves offering securities to the general investor through a registered public offering in the capital markets. Private financing involves offering unregistered securities directly to institutional investors. In a public offering, the investment banker serves as an intermediary between the issuer and purchasers and often underwrites the issue. In private financing, the investment banker serves as an agent the issuer, and the issuer sells the securities directly to the investors.

A3. Three advantages of private financing are (1) lower issuance costs, (2) faster placement for firms that do not qualify for shelf registration, and (3) greater flexibility of issue size and security arrangements. The major disadvantage of private financing is that they require a higher yield than a comparable public offering.

A4. The par value of common stock has little real significance because some states do not permit companies to sell shares at a price below par value and, consequently, par values are generally very small.

A5. Debt is by far the largest source of new external financing.

A6. Retained earnings is by far the largest source of new equity financing.

A7. Three explicit transaction cost components of the **gross underwriting spread** in an underwritten public offering are (1) the management fee--compensation for assistance in designing the issue, preparing the documentation, forming the syndicate, and directing the offering process; (2) the underwriting fee--compensation for bearing underwriting risk; and (3) the selling concession--compensation for the selling effort.

A8. The real transaction cost of an issue is the effective spread rather than the gross spread, where the former includes the market impact of the new issue. Often the market impact of the offering is larger than gross spread. The real cost of the offering should include any price changes which occur as a result of the offering as well as the explicit transaction cost.

A9. A **rights offering** is an offer to sell new stock to current shareholders on a privileged-subscription basis. The firm distributes to its shareholders rights to subscribe for additional shares at a specified price.

A10. A **dividend reinvestment plan** is a plan that gives shareholders the opportunity to automatically reinvest dividends, supplementary cash, and in some cases interest payments on the company's bonds or dividend payments on the company's preferred stock in newly issued shares of the company's common stock. The reinvested dollar amounts are paid to the firm.

A11. In a negotiated offering, the issuer selects its investment bankers, who help prepare the necessary documentation and then price and offer the issue to the public. In a competitive offering, the issuer puts the securities up for bid, and the bidding process determines which investment bankers will market the issue and at what price.

A12. **Preferred stock** is viewed as a "hybrid" security because it combines certain features of common stock and certain features of debt. It is senior to common stock but junior to debt in its claim on the firm's operating income and in its claim on the firm's assets in the event of liquidation. Preferred stock pays dividends quarterly, like common stock, but at a stated rate, like debt, and these dividends are not tax-deductible by the firm. It may have an optional redemption provision and/or a sinking fund and/or be convertible, in each case, like debt.

A13. The announcement of a public offering of common stock usually has a negative impact on the firm's share price because of the adverse selection problem involved in financing with common stock. That is, if management is operating in the best interest of shareholders, it will choose to issue new shares when it believes the shares to be relatively overvalued.

A14. a. The dividend rights associated with common stock usually require that all shares of common stock belonging to a certain class are entitled to share pro rata in any dividend distributions to that class.

b. Liquidation rights entitle common shareholders to share pro rata in any distribution of assets upon the liquidation of the firm once senior securityholders have been paid.

c. When common shares have preemptive rights, the firm must offer new shares to the existing shareholders before it offers the shares to other prospective investors.

d. Common stockholders have the right to vote, in person or by proxy, for the members of the board of directors of the firm.

B1. a. Prior to the offering, the shareholder owns 300,000 shares (= 0.30×1 million). After the offering she will own 25% of the firm (= 300,000/1.2 million), a decline of 5%.

b. To retain her ownership, she must buy 30% of the offering or 60,000 shares. Then she will still own 30% of the firm (= [300,000 + 60,000]/1.2 million).

B2. a. The stock price will probably fall, because of the adverse selection problem inherent in new issues of common. The market perceives that, when management acts in the best interests of existing shareholders, management will only offer new common stock if it believes the shares are relatively overvalued.

b. The stock market usually reacts positively to announcements of private placements. Private placements tend to reduce information asymmetries, permit better monitoring, and to provide certification of the issuer.

B3. a. Shareholder wealth has fallen by $5 million (= [29 -30]×5 million).

b. The reduction in wealth is 17% (= 5 million/29 million) of the gross proceeds of the issue.

c. Paying the underwriter costs $1.45 million (= 0.05×29 million), which is less than the $5 million decline in value resulting from the announcement.

B4. Microsoft's book value per share is $9.07 (= 5.333 billion/588 million).

B5. Exxon has 1.242 billion shares outstanding (= 1.813 billion - 571 million). The liquidation premium on Exxon's preferred stock is \$16.62 million (= 0.03×554 million). Reducing shareholders' equity by the amount of the liquidation premium results in \$37.398 billion (= 37.415 billion - 16.62 million). Thus, Exxon's book value per common share adjusted for preferred stock liquidation value is \$30.11 (= 37.398 billion/1.242 billion).

B6. If the goal of the firm is shareholder wealth maximization, then Plan II would be preferred. Under Plan II, the total value of each shareholder's position is unchanged. Before the issue, a shareholder who owned 3 shares had a total wealth of 3×(\$30/share) = \$90. After the rights offering, the same shareholder has a total wealth of \$90 (4 shares worth \$26.25 less the \$15 spent to attain the new share). Under Plan I, the value of the existing shareholders' claims is reduced due to the drop in share price. Although the explicit flotation costs are the same, the public offering includes a price impact that adversely affects existing shareholders. Note that the extra \$600,000 raised by the public offering is irrelevant.

B7. a. The number of shares sold is (70 million)/N, where N is the number of rights needed to purchase a share. In order to raise \$100 million, the subscription price must be set so that S×(70,000,000/N) = \$100 million. Solving this equation for S gives S = (10/7)N.

Finally, using Equation (23.2) with the right valued at \$0.20, the rights-on price of the stock at \$25, and the above expression for S, we have:

$$R_C = \frac{P_R - S}{N + 1} = 0.20 = \frac{25 - \frac{10}{7}\times N}{N + 1}$$

$$0.20\times N + 0.20 = 25 - \frac{10}{7}\times N; \quad 1.6286\times N = 24.8; \quad \text{so that } N = 15.228 \approx 15$$

With N = 15, S = (10/7)(15) = \$21.43.

b. Applying Equation (23.4),

$$R_C - (P_U - S)/N = (23 - 21.43)/(15) = \$0.1047$$

c. A rights offering is more appropriate than a general public offering when the firm does not have broad market appeal or has concentrated stock ownership.

d. The principal drawback to a rights offering is that a rights offering generally takes longer to complete.

B8. a. Assume that any director receiving at least 50% of the vote will win. With 20% of the firm's stock, the dissidents with certainty can elect 40% (= 0.20/0.50) of the directors, or, in this case 6 (= 0.40×15).

b. Under a noncumulative voting structure, the dissidents cannot be certain of winning any directorships.

c. At each election the dissidents can elect 40% of the directors, or 2 (= 0.40×5) per election. It would take the dissidents 3 elections to get to the 6 directorships obtained in part a.

d. Management would prefer noncumulative voting for directors because it prevents a dissident shareholder group with a relatively small percentage ownership from gaining control of directorships.

B9. a. Book value is based on historical costs and accounting measures of depreciation and generally bears no direct relationship to market value. For most firms market value exceeds book value, so selling shares at greater than book should not impact shareholder wealth. Note, too, that in a perfect capital market, selling shares above book would not change shareholder wealth.

b. The comments in part a. also apply to the case of selling stock below book.

B10. a. Preferred stock is "expensive debt" in the sense that some issues have sinking fund provisions and the fixed dividend payments are treated the same as debt interest payments by most corporations even though missing a scheduled dividend payment will not force the issuer into default. However, because preferred dividend payments are not tax-deductible, the after-tax cost of preferred will exceed the after-tax cost of debt for a taxpaying issuer.

b. Preferred stock is "cheap equity" in the sense that dividends are paid quarterly and dividend payments can be postponed without forcing the firm into default. However, because preferred stockholders have a claim on the earnings and assets of the firm that is senior to the claim of common stockholders, it has less risk and a lower required return than common stock.

B11. Due to market imperfections such as informational asymmetry, a new equity issue is generally interpreted by the market as a negative signal, and the market price of the stock will drop in response to the announcement of a new equity issue. This market impact represents a significant portion of the cost of issuing external equity, so it is important to consider this cost along with the explicit costs such as underwriter fees.

B12. a. Assume that the preferred has a par value of \$100 per share, then it is a \$2 perpetuity for which the present value is equal to the net proceeds of \$99 (= 100×[1 - 0.01]). The quarterly cost of this preferred stock is the rate, r_q, that solves the equation:

$$99 = 2/r_q;\ \text{so that, } r_q = 2.02\%, \text{ and APY } = (1.0202)^4 - 1 = 8.329\%$$

b. If the preferred stock matures in a lump sum after 10 years, the following cash flows will be associated with the issue:

Time	Item	CFBT
0	net proceeds	99
1-40	dividend payments	-2
40	principal repayment	-100

Setting the present value of the cash flows equal to the net proceeds, the quarterly cost of this preferred stock is $r_q = 2.0368\%$, corresponding to an APY = 8.399%.

c. If the preferred stock is retired in two sinking fund payments at the end of years 9 and 10, the following cash flows will be associated with the issue:

Time	Item	CFBT
0	net proceeds	99
1-36	dividend payments	-2
36	sinking fund payment	-50
37-40	dividend payments	-1
40	sinking fund payment	-50

Setting the present value of the cash flows equal to the net proceeds, the quarterly cost of this preferred stock is $r_q = 2.0381\%$, corresponding to an APY = 8.405%.

B13. Like debt, sinking fund preferred stock "matures," usually at par, according to a specified schedule. The shorter the average life of the preferred, the more debt-like it is.

B14. Using Equation (7.3),

$$r_e = 6 + 1.35\times(8) = 16.8\%$$

B15. a. Assume that the expected cash dividend of $1 is paid after the stock dividend has been paid. Thus, the expected cash dividend to a current shareholder is $1.04 (= 1×1.04). Using Equation (5.6), the cost of retained earnings is 14.16% (= 1.04/25 + 0.10).

b. The flotation-cost-adjusted current stock price is $23.75 (= 25×[1 - 0.05]). Using Equation (5.6) with the latter value gives a cost of newly issued equity of 14.38% (= 1.04/23.75 + 0.10).

B16. No information is given on the tax rate, so assume a perfect capital market with no taxes. In order to calculate the cost of equity, we need the firm's leveraged β. The leveraged β can be calculated by rearranging Equation (10.5) with $T = 0$:

$$\beta = \frac{\beta_A}{1 - L} = \frac{0.8}{1 - 0.40} = 1.33$$

The firm's cost of equity is then found with Equation (7.3).

$$r_e = 5 + 1.33\times(15 - 5) = 18.3\%$$

B17. This preferred stock is a $0.50 perpetuity for which the present value is equal to the net proceeds of $24.5625 (= 25×[1 - 0.0175]). The quarterly cost of this preferred stock is the rate, r_q, that solves the equation:

$$24.5625 = 0.50/r_q;\ \text{so that, } r_q = 2.04\%, \text{ and APY } = (1.0204)^4 - 1 = 8.39\%$$

B18. Start-up firms often raise capital more than once. By demanding preemptive rights, venture capitalists guarantee that they have the option to maintain their proportionate ownership and voting rights whenever new equity capital is raised.

B19. Rights offering are more important to large share holders who want to maintain their proportionate ownership and control. In addition, when dominant shareholders are present the selling effort can easily be focused on them.

B20. a. The rights offering protects existing shareholders from the possible wealth loss resulting from a public offering.

b. The firm in this problem has widely dispersed ownership. As a result, a rights offering would probably take a long time to complete. In addition, a rights offering eliminates the possible transactions cost savings of selling large blocks of stock to institutions.

B21. One obvious method of making a preferred issue to be like common stock is to make it convertible into common. Another method would be to create an adjustable-rate preferred whose dividend payment varied directly with the dividend payment on the common. Finally, the latter two methods could be combined.

C1. Shelf registration may lead to a more efficient capital market by increasing the firm's flexibility with respect to the timing of offerings and by reducing the issuance costs associated with individual issues.

C2. a. If only 60% of the shares were purchased at \$105 each and the remaining 40% distributed at no cost, the effective subscription price per share was \$63 (= 0.60×105).

b. Recall that N is the number of rights needed to purchase one share. Expressing N in terms of the fraction of a right distributed for each share, $N = 1/F$. Substituting for N in Equation (23.4):

$$R_C = \frac{P_E - S}{1/F} = F\times(P_E - S)$$

Substituting for R in Equation (23.3) and solving for P_E:

$$P_E = \frac{P_R + S\times F}{1 + F}$$

c. Given the unusual structure of the offering, the effective subscription price is $L\times S$ (see part a). Substituting into the expression for P_E derived in part b:

$$P_E = \frac{P_R + L\times S\times F}{1 + F}$$

d. Rearranging Equation (23.3) and substituting the expression for P_E derived in part c:

$$R_C = P_R - \frac{P_R + L\times S\times F}{1 + F}$$

e. The structure of the offering could be considered "coercive" because any shareholders who did not subscribe would see their ownership diluted by both the subscribed and unsubscribed new shares.

C3. The holder of a right (call option) distributed a by firm owns the right to buy some amount of common stock in the firm at a particular price by a particular date. The holder of a transferable put owns the right to sell stock to the firm at a particular price by a particular date. If the call/right is exercised, the holder pays the firm cash and receives stock. If the transferable put is exercised, the holder sends the firm stock and receives cash. In both cases the options are distributed proportionally. Exercising all the options would maintain the shareholder's proportional ownership. However, in one case, exercising would increase the shareholder's absolute amount of investment, in the other case it would decrease the absolute amount.

Apendix Exercises

A1. To be truly valuable, a new security must enable an investor to achieve a higher after-tax risk-adjusted return, or an issuer a lower after-tax required return, than with previously issued securities.

A2. The dividend rate on an adjustable-rate preferred is set in accord with a formula specifying a fixed margin above 3 specified Treasury yield. The dividend rate on auction-rate preferred is reset by Dutch auction every 7 weeks.

A3. The Dutch auction feature of auction-rate preferred results in the dividend rate being reset quarterly to some percentage of a market interest rate. This lowers interest-rate risk to the buyer. Remarketed preferred lessens interest-rate risk by having the dividend rate reset quarterly by a specified remarketing agent. The Dutch auction procedure has virtue of allowing the rate to be set by the market. Remarketed preferred has the virtue of great flexibility in selecting the length of the dividend period and other terms of the issue.

A4. By providing the buyer the right to convert, the issuer of convertible adjustable-rate preferred might be forced to issue new stock or use cash to by shares in the marketplace.

B1. The principal impulses to financial innovation are (1) the desire to reallocate risk and reduce yields, (2) the desire to enhance liquidity, (3) the desire to reduce agency costs, (4) the desire to reduce transaction costs, and (5) the desire to take advantage of tax arbitrage opportunities.

B2. a. The dividend rate will be set equal to the 30-year Treasury yield plus 100 basis points, or 9.7% (= 8.7 + 1).

b. The quarterly dividend during the upcoming year will be \$2.425 (= [0.097×100]/4).

Appendix Exercies continued

C1. The interest deductions in each year are computed by multiplying present value of the bond at the start of the year by the interest rate of 12%.

a. *Rounding causes small discrepancies.

Year	Present Value of the Bond	Implied Interest Deduction	Straight-line Deduction	Difference in Deductions
0	$567.43	$68.09	$86.51	$18.42
1	635.52	76.26	86.51	10.25
2	711.78	85.41	86.51	1.10
3	797.19	95.66	86.51	-9.15
4	892.86	107.14	86.51	-20.63
Total		432.56*	432.55*	0.01*

b. Based on the bond's present value of $567.43, the total interest payments over the life of the bond are $432.57 (= 1000 - 567.43). This implies that on a straight-line basis, General Electric can deduct $86.514 (= 432.57/4) in interest per bond per year.

c. The straight-line basis is more valuable because it allows General Electric to deduct larger interest payments early in the life of the bond. The value to General Electric of the larger early tax deductions can be determined by finding the present value of the differences between the deductions (displayed in the far-right column above) on an after tax basis. The source of the value is the increased present value of the tax shields.

CHAPTER 24 -- LONG-TERM DEBT

A1. Equity securities are certificates representing partial ownership in the firm while debt securities are certificates evidencing a legal obligation to make certain contractually specified payments.

A2. Unsecured debt consists of notes and debentures backed by the general credit of the firm while secured debt consists of debt securities collateralized by specific assets.

A3. A **Eurobond** is a bond issued outside the country in whose currency it is denominated.

A4. A sinking fund shortens the effective maturity of a debt issues by returning some amount of the principal prior to the stated maturity.

A5. Asymmetric information is one reason for including a call provision in a corporate bond issue. When the firm makes a new debt issue, debtholders may require a higher interest rate because of the possibility of wealth expropriation by the shareholders (for example, due to **asset substitution**). Later, if the firm has not engaged in wealth expropriation, the value of the debt may increase, making it valuable to the shareholders to call the debt. Other reasons for call provisions are tax asymmetries and transaction costs. If corporations are taxed at a higher marginal rate than bondholders, the call option is worth more to the issuer than the after-tax cost of the interest premium it must pay bondholders to obtain this privilege. Also, transaction costs associated with calling a bond issue are less than those associated with making capital market repurchases of bonds.

A6. Debt covenants are designed to protect the interests of the bondholders and to prevent the firm from taking actions that benefit its shareholders at the expense of the bondholders. Debt covenants may limit the firm's ability to incur additional indebtedness, to use cash to pay dividends or make share repurchases, to mortgage assets, or to take other actions that are detrimental to the bondholders.

A7. A **high-coupon bond refunding** occurs when the coupon on the refunded issue is greater than the coupon the market currently requires in order to value at par a new issue of bonds of the same risk class and duration.

A8. The **debt service parity** approach to evaluating the net advantage of a proposed bond refunding involves constructing (for analytical purposes) a replacement debt obligation that will have period-by-period after-tax payments that are identical to those of the existing debt obligation. If the stream of after-tax debt service payments will support sufficient new debt to cover the cost of retiring the outstanding issue, pay all transaction costs and leave a surplus, then the refinancing will provide a net benefit to the firm's shareholders (equal to the surplus) and should be undertaken.

A9. Reacquiring high-coupon debt results in a "loss" and reacquiring low-coupon debt results in a "gain" for accounting purposes. Because this reported accounting gain or loss simply reflects the difference between the immediate cost of retiring the debt to be refunded and its historical cost adjusted for any book amortization, the accounting gain or loss has no direct relationship to the impact of the refunding on shareholder wealth. The accounting treatment can give rise to an agency cost because the manager may have an incentive to refund low-coupon debt in order to realize the accounting gain even if this refunding has a negative impact on shareholder wealth.

A10. A company should not necessarily refund an outstanding debt issue the instant the net advantage of refunding first becomes positive because interest rates may continue to rise or fall and make the refunding even more profitable at a later point in time.

A11. The opportunity cost of calling bonds for refunding is zero when the market price of the bonds equals the call price.

A12. Because the firm has the option of calling the bond instantaneously, the bondholder might have to surrender the bonds for the call price at any time. As a result, in a perfect capital market, the bondholder will never pay a price (including accrued interest) higher than the price at which he might have to surrender the bonds.

A13. It may not be most advantageous for a company to call and refund an entire sinking fund debt issue immediately because the sinking fund provisions may allow the firm to redeem a portion of the issue at par and avoid having to pay the optional redemption, or call, premium on that portion of the issue that is retired through the sinking fund.

B1.

Time	Item	CFBT	CFAT	PV @ 3.05%
0	net proceeds	99,250,000	99,250,000	99,250,000
1-20	interest payments	-4,562,000	-3,011,250	-44,596,931
1-20	flotation expense	0	12,750	188,829
20	principal payment	-100,000,000	-100,000,000	-54,841,898
				PV = 0

To find the semiannual rate on your financial calculator, set PV= 99,250,000, FV = 100,000,000, PMT = 2,998,500 (= 3,011,250 - 12,750), $n = 20$, and solve for $r = 3.049$.

Time	Item	CFBT	CFAT	PV @ 6.15%
0	net proceeds	98,750,000	98,750,000	98,750,000
1-10	interest payments	-9,125,000	-6,022,500	-44,010,624
1-10	flotation expense	0	42,500	310,577
20	principal payment	-100,000,000	-100,000,000	-55,049,953
				PV = 0

To find the annual rate on your financial calculator, set PV= 98,750,000, FV = 100,000,000, PMT = 5,980,000 (= 6,022,500 - 42,500), $n = 10$, and solve for $r = 6.151$.

B2. We repeat the analysis in Table 24-3 with a 5.2% semiannual coupon rate for the installment debt:

	AT Debt Service	Savings/Cost
Refunding Issues:		
Conventional (5.2%)	\$3,420,667	\$100,000,000
Installment (5.2%)	530,833	9,846,817
Total	\$3,951,500	\$109,846,817
Refunded Issue (6%)	\$3,951,500	100,000,000
Present Value Savings		\$9,846,817
Total AT Transaction Costs		\$5,776,000
Net Advantage of Refunding		\$4,070,817

B3. The pretax semiannual cash flows associated with the 5-year life of the issue are:

Time	Item	CFBT
0	net proceeds	950
1-8	interest payments	-50
8	sinking fund payment	-500
9-10	interest payments	-25
10	sinking fund payment	-500

The semiannual pretax cost of debt is the rate, *r*, that solves the following equation:

$$950 = \sum_{t=1}^{7} \frac{50}{(1+r)^t} + \frac{550}{(1+r)^8} + \frac{25}{(1+r)^9} + \frac{525}{(1+r)^{10}}$$

If your calculator has uneven cash flow capabilities, you can solve for *r* directly; otherwise, use trial and error to find $r = 5.728\%$. Then the annual pretax cost of debt is APY = $(1.05728)^2 - 1 = 11.78\%$.

B4. a. As in the previous problem, we need to identify the semiannual cash flows (this time the after-tax cash flows) associated with the issue:

Time	Item	CFBT	CFAT
0	net proceeds	980	980.00
1-20	interest payments	-60	-30.00
1-20	flotation expense	0	0.50
20	principal repayment	-1000	-1000.00

Solving for the interest rate that makes the present value of the after-tax cash flows equal to the net proceeds, we find that the semiannual after-tax cost of debt is 3.085%. The annual after-tax cost of debt is: APY = $(1.03085)^2 - 1 = 6.27\%$

b. If the firm is not a taxpayer and expects never to be one, the appropriate cash flows are the above CFBTs. Setting the present value of the before-tax cash flows equal to the net proceeds, the semiannual after-tax cost of debt is 6.177% and the APY = $(1.06177)^2 - 1 = 12.74\%$

c. If the firm expects to begin paying taxes at a 50% rate after 5 years (that is, beginning in the 6th year), the after-tax cash flows would be as follows:

Time	Item	CFBT	CFAT
0	net proceeds	980	980.00
1-10	interest payments	-60	-60.00
1-10	flotation expense	0	1.00
11-20	interest payments	-60	-30.00
11-20	flotation expense	0	0.50
20	principal repayment	-1000	-1000.00

Setting the present value of the after-tax cash flows equal to the net proceeds, the semiannual after-tax cost of debt is 4.933%, and the after-tax APY $= (1.04933)^2 - 1 = 10.11\%$

The preceding calculation assumes that the firm loses the interest and flotation expense tax deduction for years 1-5 forever. If the firm can carry them forward (which is permitted in many cases under the Internal Revenue Code) and use them fully in year 6, the period 11 CFAT is $-30.00 + 0.50 + 10\times(0.5)\times(60) + 10\times(0.5)\times(1.0) = 275.5$. Setting the present value of the after-tax cash flows equal to the net proceeds, the semiannual after-tax cost of debt is 3.329%, and the APY $= (1.03329)^2 - 1 = 6.77\%$

B5. a.

Time	Item	CFBT
0	Proceeds	+1000.0
1	Interest	-100.0
2	Interest	-102.5
3	Interest	-105.0
4	Interest	-107.5
5	Interest	-110.0
5	Principal	-1000.0

b. Setting the present value of the before-tax cash flows equal to the proceeds from the issue, the pre-tax cost of debt is the rate is $r = 10.45\%$.

c. Assuming a 34% tax rate, we can calculate the following after-tax cash flows:

Time	Item	CFBT	CFAT
0	Proceeds	+1000.0	+1000.00
1	Interest	-100.0	-66.00
2	Interest	-102.5	-67.65
3	Interest	-105.0	-69.30
4	Interest	-107.5	-70.95
5	Interest	-110.0	-72.60
5	Principal	-1000.0	-1000.00

Setting the present value of the after-tax cash flows equal to the proceeds from the issue, the after-tax cost of debt is $r = 6.91\%$.

B6.

	U.S. Public Offering	U.S. Private Offering	Eurobond Offering
Maturity	7 years	7 years	7 years
Interest Rate	8% per annum	8 3/8% per annum	8 1/8% per annum
Coupon Frequency	Semiannual	Semiannual	Annual
Flotation Costs	$900,000	$500,000	$1,100,000

a. For the U.S. Public offering we have the following cash flows:

Time	Item	CFBT	CFAT
0	net proceeds	99,100,000	99,100,000
1-14	interest payments	-4,000,000	-2,640,000
1-14	flotation expense	0	10,929
14	principal repayment	-100,000,000	-100,000,000

To find the semiannual rate on your financial calculator, set PV= 99,100,000, FV = 100,000,000, PMT = 2,629,071 (= 2,640,000 - 10,929), n = 14, and solve for r = 2.707.

For the U.S. Private offering we have the following cash flows:

Time	Item	CFBT	CFAT
0	net proceeds	99,500,000	99,500,000
1-14	interest payments	-4,187,500	-2,763,750
1-14	flotation expense	0	6071
14	principal repayment	-100,000,000	-100,000,000

To find the semiannual rate on your financial calculator, set PV= 99,500,000, FV = 100,000,000, PMT = 2,757,679 (= 2,763,750 - 6071), n = 14, and solve for r = 2.80.

For the Eurobond offering we have the following cash flows:

Time	Item	CFBT	CFAT
0	net proceeds	98,900,000	98,900,000
1-14	interest payments	-8,125,000	-5,362,500
1-14	flotation expense	0	53,429
14	principal repayment	-100,000,000	-100,000,000

[Put in PV= 98,900,000, FV = 100,000,000, PMT = 5,309,071 (= 5,362,500 - 53,429), n = 7, and solve for r = 5.50. Then solve for the semiannual rate: $(1.055)^{0.50} - 1 = 2.71\%$.

b. The U.S. Public offering's semiannual rate is slightly lower (2.707%) than the Eurobond offering's semiannual rate (2.71%).

c. Because the U.S. Public offering rate and the Eurobond rate are so close, Exxon would probably base its decision as to which to offer on other factors. These factors would include: which issue might need to have more restrictive covenants; whether any international tax law changes were likely; and the difficulty of refunding Eurobonds due to their being bearer bonds.

B7. Cumulative dividends cannot exceed $115 million (= 25 + 0.60×[3]×[50]). Dividends paid to date = $60 million. Barring any other covenants, the firm can pay a dividend of $55 million (= 115 - 60).

B8. With $100 million in EBIT, the firm can have total interest payments of $40 million (= 100/2.5), which is an increase of $15 million over current payments. Thus, it can issue $150 million (= 15/0.10) in new debt.

B9. A limitation on sale and-leaseback require a borrower to meet certain financial tests before selling an asset then leasing it back, thereby receiving cash and contracting to make lease payments. A limitation on liens or a negative pledge clause requires a borrower to meet certain financial tests before mortgaging an asset, thereby receiving cash and contracting to make mortgage payments. Both types of limitation prohibit a borrower from "selling" an asset to raise funds while incurring an obligation to make fixed payments.

B10. We need to find that number (call it x) when added to both total assets and long-term debt will result in an asset-coverage ratio of 1.5. Solving $(1 + x)/(0.40 + x) = 1.5$ for $x = 0.80$. The firm can issue an additional $800 million in long-term debt.

B11. Applying Equation (24.1), with each weight = 0.1429 (= 1/7):

Average Life = 0.1429×(14) + 0.1429×(15) + ... + 0.1429×(20) = 17 years

B12. Using Equation (24.1), the issue's average life is 8.8 years = 0.2×(7) + 0.2×(8) + 0.2×(9) + 0.4×(10)

B13. The bond issue has 16 periods to maturity. In the first eleven periods and in periods 13 and 15, it pays interest only of $2.1 million. At the end of periods 12, 14, and 16 it pays interest plus principal of $20 million.

a. The bond's price is:

$$P = \sum_{t=1}^{16} \frac{CF_t}{(1.03)^t} = \$66.09 \text{ million}$$

b. The bond's average life is 7 years (= 0.333×[6] + 0.333×[7] + 0.333×[8]).

c. Using Equation (24.4), the bond's duration is:

$$\text{Duration} = \frac{1}{66.09} \times \left[\sum_{t=1}^{16} \frac{t \times CF_t}{(1.03)^t}\right] = 11.51 \text{ periods}$$

Duration = 5.75 years (= 11.51/2).

d. Duration provides a better measure of the average timing of the firm's total cash flow, because it incorporates all the bond's cash flows.

B14. a. The bond's semiannual rate is found by solving the following equation f

$$1050 = \sum_{t=1} \frac{50}{(1+r)^t} + \frac{333.33}{(1+r)^{16}} + \frac{333.33}{(1+r)^{18}} + \frac{333.33}{(1+r)^{20}}$$

$r = 4.9\%$ semiannually, so the bond's yield to maturity is 9.8% (= 4.9×[2

To solve the above equation for r must be done by trial-and-error, computer, or financial calculator. One way to solve it with a calculator would be to use the cash flow features of the calculator. Enter the relevant cash flows for each year and then compute IRR = $r/2$.

b. The bond's average life is 9 years (= 0.333×[8] + 0.333×[9] + 0.333×[10]).

c. Using Equation (24.4), the bond's duration is:

$$\text{Duration} = \frac{1}{1050} \times \left[\sum_{t=1}^{20} \frac{t \times CF_t}{(1.049)^t} \right] = 16.78 \text{ periods}$$

Duration = 8.39 years (= 16.78/2).

B15. a. The period-by-period after-tax debt service on the 15% debt issue is:

Time	Item	CFBT	CFAT
1-20	Interest	3,750,000	2,250,000
1-20	Amort. Expense	-	-16,000
1-20	Net		2,234,000
20	Principal payment	50,000,000	50,000,000

b. The after-tax interest payment on the new issue would be $1,875,000 (= [1 - 0.4]×[0.0625]× 50 million). The tax shield due to amortization of the issue expense on the new issue would be $20,000 (= 0. 4×[1,000,000/20]). Thus, the net semiannual after-tax payment on the new issue would be $1,855,000 (= 1,875,000 - 20,000). The semiannual after-tax savings is the difference between the semiannual after-tax payment on the old issue and the semiannual after-tax payment on the new issue, which is $379,000 (= 2,234,000 - 1,855,000).

c. The amount of installment debt the company can issue is the present value of the $379,000 savings over 10 years (20 semiannual periods), discounted at the after-tax cost of new installment debt, 3.6% (= [1 - 0.4]× 6%). This present value is $5,338,086.

d. The net advantage of refunding the 15% debt issue is the difference between the amount of installment debt that can be issued and the total after-tax transaction costs:

Transaction Costs:	
After-tax Redemption Premium	$3,000,000
Write-off of Unamortized Discount	-320,000
New Issue Expenses	1,000,000
Opportunity Cost	1,000,000
Total	$4,680,000

Net Advantage = 5,338,086 - 4,680,000 = $658,086

. a. We compute the net advantage by applying Equation (24.4) and assuming $r^* = r'$:

$$NA = \sum_{n=1}^{N} \frac{(1-T)\times(r-r')\times D + T\times([E-U]/N)}{[1+(1-T)\times r^*]^n} - (1-T)\times(P-D) - [E+(1-T)\times F - T\times U] - (P-B)$$

$$= \sum_{n=1}^{40} \frac{(1-0.4)\times(0.06-0.05)\times 100M + 00.4\times([750{,}000-500{,}000]/40)}{[1+(1-0.4)\times 0.05]^n} - (1-0.4)\times(2M)$$

$$- [750{,}000 + (1-0.4)\times 300{,}000 - 00.4\times(500{,}000)] = \$11{,}996{,}650$$

b. If the bonds are selling at $1017, then B is not equal to P and the opportunity cost of calling the bonds at a price above the market value reduces the net advantage of the refunding. Repeating part a with $B = \$101{,}700{,}000$ gives a net advantage of $11,696,650.

B17. With a flat term structure, $r^* = r'$, and we can apply Equation (24.4) to compute the net advantage of refunding:

$$NA = \sum_{n=1}^{N} \frac{(1-T)\times(r-r')\times D + T\times([E-U]/N)}{[1+(1-T)\times r^*]^n} - (1-T)\times(P-D) - [E+(1-T)\times F - T\times U] - (P-B)$$

$$= \sum_{n=1}^{20} \frac{(1-0.4)\times(0.06-0.055)\times 70M + 0.4\times([500{,}000-350{,}000]/20)}{[1+(1-0.4)\times 0.055]^n} - (1-0.4)\times(77M-70M)$$

$$- [500{,}000 - 0.4\times(350{,}000)] - (77M - 74.2M) = -\$4{,}277{,}235$$

B18. We can apply Equation (24.4), assuming that $r^* = r'$ and then deduct the cost of the overlapping interest. The net cost of the overlapping interest is the difference between the interest cost on the old bonds and the interest earned on the short-term investment in marketable securities net of taxes:

$$\text{Net cost} = (1 - 0.4)\times[50{,}000{,}000\times(0.15)\times(1/12) - 50{,}000{,}000\times(0.08)\times(1/12)] = 0.6\times[625{,}000 - 333{,}333] = \$175{,}000$$

Applying Equation (24.4) gives:

$$NA = \sum_{n=1}^{N} \frac{(1-T)\times(r-r')\times D + T\times([E-U]/N)}{[1+(1-T)\times r^*]^n} - (1-T)\times(P-D) - [E+(1-T)\times F - T\times U] - (P-B)$$

$$= \sum_{n=1}^{30} \frac{(1-0.4)\times(0.075-0.055)\times 50M + 0.4\times([400{,}000-300{,}000]/30)}{[1+(1-0.4)\times 0.055]^n} - (1-0.4)\times(52.5M-50M)$$

$$- [400{,}000 - 0.4\times(300{,}000)] - (52.5M - 52.75M) = \$9{,}812{,}171$$

Finally, deducting the net cost of the overlapping interest gives the net advantage of refunding the issue:

$$NA = 9{,}812{,}171 - 175{,}000 = \$9{,}637{,}171$$

B19. We can apply Equation (24.4), assuming that $r^* = r'$, and then deduct the cost of the overlapping interest. In this case, however, the earnings on short-term marketable securities exactly offsets the cost of interest on the old bonds, so the net advantage is computed directly from Equation (24.4) by noting that the appropriate issue costs to consider are the issue costs for the portion of the new issue that will be used to retire the old debt, $E = \$192{,}000$ (= 480,000 ×[20/50]):

$$NA = \sum_{n=1}^{52} \frac{(1-0.35)\times(0.05-0.046)\times 20M + 0.35\times([192{,}000-185{,}000]/52)}{[1+(1-0.35)\times 0.046]^n} - (1-0.35)\times(20.4M-20M)$$

$$- [192{,}000-0.35\times(185{,}000)] - (20.4M-20.6M) = \$1{,}177{,}285$$

B20. Applying the "back of the envelope estimate" of Equation (24.4), with E = F = U = 0, the net advantage of refunding an amount representing the first sinking fund payment is:

$$\text{NA} = \sum_{n=1}^{2} \frac{(1-0.34)\times(0.07-0.06)\times(2M)}{[1+(1-0.34)\times(0.06)]^n} - (1-0.34)\times(0.02)\times(2M) = -\$1489$$

The net advantage of refunding the second sinking fund payment is:

$$\text{NA} = \sum_{n=1}^{4} \frac{0.66\times(0.01)\times(2M)}{[1+0.66\times(0.06)]^n} - 0.66\times(0.02)\times(2M) = \$21{,}560$$

The net advantage of refunding the third sinking fund payment is:

$$\text{NA} = \sum_{n=1}^{6} \frac{(1-0.34)\times(0.07-0.065)\times(2M)}{[1+(1-0.34)\times 0.065]^n} - (1-0.34)\times(0.02)\times(2M) = \$7874$$

The net advantage of refunding the fourth sinking fund payment is:

$$\text{NA} = \sum_{n=1}^{8} \frac{(1-0.34)\times(0.005)\times(2M)}{[1+(1-0.34)\times 0.065]^n} - (0.66)\times(0.02)\times(2M) = \$17{,}509$$

The net advantage of refunding the fifth sinking fund payment is:

$$\text{NA} = \sum_{n=1}^{10} \frac{(1-0.34)\times(0.005)\times(2M)}{[1+(1-0.34)\times 0.065]^n} - (0.66)\times(0.02)\times(2M) = \$26{,}367$$

Southern should call \$8 million of the issue for immediate redemption and should apply that amount to the second through fifth sinking fund payment obligations. Southern should wait one year to redeem the remaining \$2 million at par through the ordinary operation of the sinking fund.

B21. a. Applying the back of the envelope version of Equation (24.4) used in B20 above, the net advantage of refunding an amount representing the first sinking fund payment is:

$$\text{NA} = \frac{(1-0.34)\times(0.10-0.08)\times(20M)}{[1+(1-0.34)\times(0.08)]} - (1-0.34)\times(0.03)\times(20M) = -\$145{,}240$$

The net advantage of refunding the second sinking fund payment is:

$$NA = \frac{0.66\times(0.02)\times(20M)}{1+0.66\times(0.08)} + \frac{0.66\times(0.02)\times(20M)}{(1+0.66\times(0.08))^2} - 0.66\times(0.03)\times(20M) = \$92,944$$

The net advantage of refunding the third sinking fund payment is:

$$NA = \frac{0.66\times(0.02)\times(20M)}{1+0.66\times(0.08)} + \frac{0.66\times(0.02)\times(20M)}{(1+0.66\times(0.08))^2} + \frac{0.66\times(0.02)\times(20M)}{(1+0.66\times(0.08))^3} - 0.66\times(0.03)\times(20M)$$
$$= \$319,182$$

b. The company should call $40 million for immediate redemption and apply that amount to meet the second and third sinking fund payment obligations. It should wait 1 year to redeem the other $20 million at par through the normal operation of the sinking fund.

B22. Applying the "back of the envelope estimate" of Equation (24.4), with E = F = U =0, the net advantage of refunding an amount representing the first sinking fund payment is:

$$NA = \sum_{n=1}^{20} \frac{(1-0.4)\times(0.05-0.04)\times(50M)}{[1+(1-0.4)\times(0.04)]^n} - (1-0.4)\times(0.03)\times(50M) = \$3,325,398$$

The net advantage of refunding the second sinking fund payment is:

$$NA = \sum_{n=1}^{22} \frac{0.6\times(0.05-0.0425)\times(50M)}{[1+0.6\times(0.0425)]^n} - 0.6\times(0.03)\times(50M) = \$2,852,949$$

The net advantage of refunding the third sinking fund payment is:

$$NA = \sum_{n=1}^{24} \frac{(0.6)\times(0.05-0.045)\times(50M)}{[1+(0.6)\times 0.045]^n} - (0.6)\times(0.03)\times(50M) = \$2,002,430$$

Ford should refund the entire issue.

B23. No, a firm should not necessarily refund an outstanding debt issue the instant the net advantage of refunding first becomes positive because interest rates may continue to rise or fall and make the refunding even more profitable at a later point in time.

C1. a. If the debt remains outstanding until maturity, the issue's cash flows will be:

Time	Item	CFBT	CFAT
0	net proceeds	990	990.00
1-20	interest payments	-60	-39.60
1-20	flotation expense	0	0.17
20	principal repayment	-1000	-1000.00

We find the semiannual after-tax cost of debt by setting the present value of the after-tax cash flows equal to the net proceeds. In this case, the after-tax rate that solves the equation is 4.017%, which corresponds to an after-tax APY cost of debt of 8.2%.

b. If the investors put the bonds back to the firm at the end of the fifth year, the cash flows associated with the issue are the following:

Time	Item	CFBT	CFAT
0	net proceeds	990	990.00
1-10	interest payments	-60	-39.60
1-10	flotation expense	0	0.17
10	redemption payment	-1030	-1018.10

The $30 redemption premium and $5 in unamortized issuance expense are tax-deductible and reduce taxes by $11.9 (= 0.34×[35]), so the terminal payment is $1018.10 (= 1030 - 11.90).

We find the semiannual after-tax cost of debt by setting the present value of the after-tax cash flows equal to the net proceeds. With the redemption, the after-tax rate that solves the equation is 4.217%, which corresponds to an after-tax APY cost of debt of 8.61%.

C2. a. We apply Equation (24.4) to the $190 million portion of the issue that was tendered; as a result, we have the following variables:

B = 1,120/bond×(190,000 bonds) = $212,800,000
D = 1000/bond×(190,000 bonds) = $190,000,000
E = 15/bond×(190,000 bonds) = $2,850,000
F = 3/bond×(190,000 bonds) = $570,000
N = 37×(2) = 74 semiannual periods
P = 1145/bond×(190,000 bonds) = $217,550,000
r = 5.8125% (= 11.625/2) per semiannual period
$r' = r^*$ = 4.5% (= 9/2) per semiannual period
T = 0.34
U = $6,591,100 (= [1000 - 965.31]×190,000)

$$NA = \sum_{n=1}^{N} \frac{(1-T)\times(r-r')\times D + T\times([E-U]/N)}{[1+(1-T)\times r^*]^n} - (1-T)\times(P-D) - [E+(1-T)\times F - T\times U] - (P-B)$$

$$= \sum_{n=1}^{74} \frac{(1-0.34)\times(0.058125-0.045)\times 190M + 0.34\times([2.85M-6.5911M]/74)}{[1+(1-0.34)\times 0.045]^n} - (1-0.34)$$

$$\times(217.55M-190M) - [2.85M+(1-0.34)\times 0.57M - 0.34\times(6.5911M)] - (217.55M-212.8M)$$

$$= \sum_{n=1}^{74} \frac{1{,}628{,}686}{(1.0297)^n} - 23{,}918{,}226 = \$24{,}632{,}039$$

b. We again apply Equation (24.4), but this time to the entire $250 million issue. We also assume that the market price of the bonds at the time of the call will be equal to the call price, $1075. As a result, we have the following variables:

$B = P$ = 1075/bond×(250,000 bonds) = $268,750,000
D = 1000/bond×(250,000 bonds) = $250,000,000
E = 15/bond×(250,000 bonds) = $3,750,000
F = 3/bond×(250,000 bonds) = $750,000
N = 35×(2) = 70 semiannual periods
r = 5.8125% (= 11.625/2) per semiannual period
$r' = r^*$ = 4.5% (= 9/2) per semiannual period
T = 0.34
U = $8,202,500 (= [1000 - 967.19]×250,000)

$$NA = \sum_{n=1}^{70} \frac{(1-0.34)\times(0.058125-0.045)\times 250M + 0.34\times([3.75M-8.2025M]/70)}{[1+(1-0.34)\times 0.045]^n}$$

$$- (1-0.34)\times(268.75M-250M) - [3.75M+(1-0.34)\times 0.75M-0.34\times(8.2025M)]$$

$$= \sum_{n=1}^{70} \frac{2{,}143{,}999}{(1.0297)^n} - 13{,}831{,}150 \ = \ \$49{,}052{,}329$$

This net advantage would be realized two years hence. To find the net advantage at time zero, the value just calculated is discounted back four periods at a rate of 2.97% per period:

$$NA = 49{,}052{,}329/(1.0297)^4 = \$43{,}633{,}172$$

c. Mountain States could have used a break-even analysis or dynamic programming and the calculation of a stopping curve to determine the optimal timing for the debt refunding.

C3. a. This debt is premium rather than the usual discount debt. So we need to modify Equation (24.4) because the old debt is retired at a discount to its par value and that discount may be a taxable gain. The basic DCF approach still applies: Discount the change in after-tax debt service at the after-tax cost of the new debt issue. Four changes are required to Equation (24.4).

(1) The firm can retire D of the debt for a price of $B < D$. If it issues B of the new debt to pay this cost and the new debt matures at the same time the old debt was scheduled to, the firm saves $D - B$ at maturity, time N. The present value of this amount, calculated at the after tax cost of the new issue $(1 - T)r'$, is $[D - B]/[1 + (1 - T)r']^N$.

(2)The old debt has interest expense of rD per bond, and the new cebt has interest expense of $r'B$ per old bond. The change in after-tax interest expense is then $(1 - T)(rD - r'B)$, the present value of which is also calculated by discounting at $(1 - T)r'$. This term replaces the present value of $(1 - T)(r - r')D$ in Equation (24.4).

(3) The firm might have to pay income tax on the difference $D - B$. (It's treated as "forgiveness of debt" income under the tax code.) If the tax rqte is T' (which migth differ from T, or it could be the same as T), then the tax liability (if any) is $T'(D - B)$. Since the firm retires the debt at a discount, and not at a premium, this term replaces $(1 - T)(P - D)$ in Equation (24.4).

(4) Since the firm pays the market price of the bonds to buy them, the last term, $P - B$, is zero in this case.

continued . . .

C3. cont.

With these changes, Equation (24.4) becomes

$$\text{NA} = \sum_{n=1}^{N} \frac{(1-T)\times(rD-r'B)+T\times([E-U]/N)}{[1+(1-T)\times r']^n} + \frac{D-B}{[1+(1-T)\times r']^N} - T'\times(D-B)-[E+(1-T)\times F-T\times U]$$

Because the gain is nontaxable, set $T' = 0$ in the preceding equation, and NA = \$475,193:

$$\text{NA} = \sum_{n=1}^{N} \frac{(1-T)\times(rD-r'B)+T\times([E-U]/N)}{[1+(1-T)\times r']^n} + \frac{D-B}{[1+(1-T)\times r']^N} - T'\times(D-B)-[E+(1-T)\times F-T\times U]$$

$$= \sum_{n=1}^{20} \frac{(1-0.34)\times[0.05\times(50M)-0.075\times(38.8525M)]+0.34\times([0.38M-0.40M]/20)}{[1+(1-0.34)\times 0.075]^n}$$

$$+ \frac{50M-38.8525M}{[1+(1-0.34)\times 0.075]^{20}} -[0.38M+(1-0.34)\times 0.15M-0.34\times(0.40M)]$$

$$= -3{,}423{,}396 + 4{,}241{,}589 - 343{,}000 = \$475{,}193$$

b. With the gain immediately taxable at a 34% rate, we can use the same modified equation with $T' = 0.34$ to compute NA = -\$3,314,957:

$$\text{NA} = \sum_{n=1}^{N} \frac{(1-T)\times(r\times D-r'\times B)+T\times([E-U]/N)}{[1+(1-T)\times r']^n} + \frac{D-B}{[1+(1-T)\times r']^N} - T'\times(D-B)-[E+(1-T)\times F-T\times U]$$

$$= \sum_{n=1}^{20} \frac{(1-0.34)\times[0.05\times(50M)-0.075\times(38.8525M)]+0.34\times([0.38M-0.40M]/20)}{[1+(1-0.34)\times 0.075]^n}$$

$$+ \frac{50M-38.8525M}{[1+(1-0.34)\times 0.075]^{20}} - 0.34\times(50M-38.8525M) - [0.38M+(1-0.34)\times 0.15M-0.34\times(0.40)M]$$

$$= -3{,}423{,}396 + 4{,}241{,}589 - 3{,}790{,}150 - 343{,}000 = -\$3{,}314{,}957$$

c. Finally, with the gain taxable at a 34% rate at the time the issue would have matured (10 years from now), the *present value tax rate* on the gain is the present value of 34% discounted for 20 periods at 4.95%. As a result, $T' = 0.129369$ and NA = -\$966,948:

$$\text{NA} = \sum_{n=1}^{20} \frac{(1-0.34)\times[0.05\times(50M)-0.075\times(38.8525M)]+0.34\times([0.38M-0.40M]/20)}{[1+(1-0.34)\times 0.075]^n}$$

$$+ \frac{50M-38.8525M}{[1+(1-0.34)\times 0.075]^{20}} - 0.129369\times(50M-38.8525M) - [0.38M+(1-0.34)\times 0.15M-0.34\times(0.40M$$

$$= -3{,}423{,}396 + 4{,}241{,}589 - 1{,}442{,}141 - 343{,}000 = -\$966{,}948$$

d. This refunding is only profitable in the case where the gain is nontaxable.

C4. a. If the outstanding bonds have a 12% yield, their market price is the present value of $30 per period of interest payments for 12 periods and the $1000 principal repayment at the end of the 12 semiannual periods, all discounted at 6% per period. This present value is $748.48 per bond, for an aggregate market value of $74.848 million.

b. With a nontaxable gain, we can use the modified Equation (24.4), which we developed in the solution to the previous problem, C3, with $T' = 0$. NA = $5,590,269:

$$NA = \sum_{n=1}^{12} \frac{(1-0.34)\times[0.03\times(100M)]-0.06\times(74.848M)+0.34\times([1.2M-0.60M]/12)}{[1+(1-0.34)\times0.06]^n}$$

$$+\frac{100M-74.848M}{[1+(1-0.34)\times0.06]^{12}} - [1.2M+0.1M-0.34\times(0.60M)]$$

$$= -9{,}096{,}285 + 15{,}782{,}554 - 1{,}096{,}000 = \$5{,}590{,}269$$

c. With the gain taxable at the end of 6 years at a 34% rate, the *present value tax rate* on the gain is $T' = 0.2133456$ ($= 0.34/(1.0396)^{12}$), and NA = $224,200:

$$NA = \sum_{n=1}^{12} \frac{(1-0.34)\times[0.03\times(100M)-0.06\times(74.848M)]+0.34\times([1.2M-0.60M]/12)}{[1+(1-0.34)\times0.06]^n}$$

$$+\frac{100M-74.848M}{[1+(1-0.34)\times0.06]^{12}} - 0.2133456\times(100M-74.848M) - [1.2M+0.1M-0.34\times(0.60M)]$$

$$= -9{,}096{,}285 + 15{,}782{,}554 - 5{,}366{,}069 - 1{,}096{,}000 = \$224{,}200$$

d. With the gain immediately taxable at a 34% rate, $T' = 0.34$ in the same modified equation, and NA = -$2,961,411:

$$NA = \sum_{n=1}^{12} \frac{(1-0.34)\times[0.03\times(100M)-0.06\times(74.848M)]+0.34\times([1.2M-0.60M]/12)}{[1+(1-0.34)\times0.06]^n}$$

$$+\frac{100M-74.848M}{[1+(1-0.34)\times0.06]^{12}} - 0.34\times(100M-74.848M) - [1.2M+0.1M-0.34\times(0.60M)]$$

$$= -9{,}096{,}285 + 15{,}782{,}554 - 8{,}551{,}680 - 1{,}096{,}000 = -\$2{,}961{,}411$$

C5. a. We can modify Equation (24.4) to account for the fact that the preferred dividends, redemption premium, and issue expenses are not deductible for tax purposes. With P = $55 million and D = $50 million, we have:

$$NA = \sum_{n=1}^{80} \frac{(0.04-0.03)\times50M}{(1.03)^n} - (55M-50M)-1M = \$9{,}100{,}382$$

b. With perpetual preferred stock, the sum is replaced by the present value of the perpetuity of dividend savings:

$$NA = \frac{(0.04-0.03)\times50M}{0.03} - (55M-50M)-1M = \$10{,}666{,}667$$

C6. a. Start with Equation (24.4). Preferred stock dividends are paid quarterly, and neither the dividends nor the issuance expenses nor the redemption premium are tax-deductible. Also, the dividend savings are realized in per perpetuity so that $r^* = r'$. Setting $T = 0$ and $r^* = r'$ in Equation (24.4) gives

$$NA = \sum_{n=1}^{\infty} \frac{(r-r')D}{(1+r')^n} - (P-D) - E - (P-B) = \frac{(r-r')}{r'}D - (P-D) - E - (P-B)$$

where
B = the market price of the refunded (that is, old) preferred stock issue;
D = the par value of the old issue;
E = new issue expenses and other costs of the refunding;
P = the call (strike) price of the old issue (excluding accrued dividends);
r = the dividend rate on the old issue; and
r' = the dividend rate on the new issue.

b. The break-even refunding rate, b, gives a net advantage equal to zero. Substituting b for r' in the expression developed in part a and setting that expression equal to zero gives

$$NA = 0 = \frac{(r-b)}{b} \times D - [(P-D) + E + (P-B)]$$

Solving for b gives

$$b = \frac{r \times D}{P + E + (P-B)}$$

C7. a. The after-tax cost of the 2-year issue is 7.92% (= [1 - 0.34]× 12). The break-even refunding rate is the rate, r, that solves the following equation:
so that, r = 14.16%

$$1000 = \frac{(1-0.34)\times(0.10)\times 1000}{1.0792} + \frac{(1-0.34)\times 1{,}000 \times r + 1000}{(1.0792)^2}$$

b. If the firm believes it can refund the 1-year issue at its maturity for a rate less than 14.16%, it should issue the 1-year debt. Otherwise, it should issue the 2-year debt.

c. The issuer's actual realized cost of 2-year funds is the rate, r_2, that solves the following equation:

$$1000 = \frac{(1-0.34)\times(0.10)\times 1000}{1 + (1-0.34)\times r_2} + \frac{(1-0.34)\times(0.13)\times 1000 + 1000}{[1 + (1-0.34)\times r_2]^2}$$

so that, r_2 = 11.45% before tax; and (1 - 0.34)× r_2 = 7.56% after tax

d. The sequence of two 1-year debt issues should be accepted if the firm knew with certainty that the 1-year new issue rate 1 year hence would be 13%.

CHAPTER 25 -- LEASING AND OTHER ASSET-BASED FINANCING TECHNIQUES

A1. A **lease** is a rental agreement that extends over a specified term of 1 year or longer under which the owner of an asset grants another party the exclusive right to use the asset in return for a series of fixed payments. The **lessor** is the owner of the asset and the **lessee** is the party that pays for the right to use the asset. The lessor and the lessee represent the two sides of a lease transaction.

A2. Under a **direct lease**, the lessee either leases the asset directly from the manufacturer or arranges for some other lessor to buy it from the manufacturer and lease it to the lessee. Under a **sale-and-leaseback** agreement, the owner of an asset sells it for cash; the purchaser assumes legal ownership and then leases the asset back to the original owner. Under a **leveraged lease** arrangement, the lessor borrows a substantial portion of the purchase price of the asset and secures the loan by mortgaging the asset and assigning the lease contract to the lender.

A3. The principal advantages of lease financing are (1) a more efficient utilization of tax deductions and tax credits of ownership, (2) reduced risk, (3) reduced cost of borrowing, and (4) the ability to circumvent restrictive debt covenants or other restrictions. The principal disadvantages of lease financing are the lessee's forfeiture of the tax deductions associated with asset ownership and, in most cases, the loss of residual value. Two purported advantages that are really of dubious value are (1) off-balance-sheet financing to minimize the impact of the financing on the firm's apparent degree of leverage and (2) the preservation of capital by avoiding bank borrowing or the investment of internally generated funds.

A4. Leasing reduces the firm's debt capacity because the firm must meet its lease payment obligations in a timely manner in order to have uninterrupted use of the asset. The consequences of failing to make a lease payment are the same as the consequences of failing to pay interest or repay principal on outstanding debt. For the firm's capital structure to remain unchanged under a lease arrangement, the firm must reduce its existing debt commensurately, so that the lease displaces conventional debt financing dollar-for-dollar.

A5. The principal tax benefits associated with asset ownership are the accelerated depreciation deductions and any tax credits, such as investment tax credits, that may be available.

A6. To qualify as a **true lease** for tax purposes, the lease must satisfy the following requirements: (1) the term of the lease cannot exceed 80% of the useful life of the asset; (2) the lessor must maintain a minimum equity investment in the asset throughout the term of the lease of no less than 10% of the asset's original cost; (3) the lessor can grant the lessee a purchase option only if the option price is equal to the asset's fair market value at the time the purchase option is exercised; (4) the lessee does not pay any portion of the purchase price of the asset and, if the lease is a leveraged lease, does not lend the lessor funds with which to purchase the leased asset or guarantee loans from others to the lessor for this purpose; and (5) the lessor must hold title to the property and must represent and demonstrate that it expects to earn a pretax profit from the lease transaction.

A7. The **net advantage of leasing** is measured as the difference between the purchase price and the present value of the incremental after-tax cash flows associated with the lease.

A8. The appropriate discount rate to use in calculating the present value of the incremental after-tax cash flows associated with a lease financing is the lessee's after-tax cost of similarly secured debt, assuming 100% debt financing for the asset. The expected residual value of the asset is discounted at a higher rate because the residual value is more closely related to overall project economic risk than to financing risk; while the lease payments are contractually specified like debt interest and principal payments, the residual value depends on the riskiness of the project.

A9. a. First, we compute the new before-tax cost of debt and then apply Equation (25.2):

$$r' = 0.8\times(11\%) + 0.2\times(14\%) = 11.6\%$$

$$\text{NAL} = 10M - \sum_{t=1}^{10} \frac{(1-0.4)\times 1.7M + 0.38M}{[1+(1-0.4)\times 0.116]^t} - \frac{0.5M}{[1+0.14]^{10}} = \$13{,}907$$

NACCO should lease in this case.

b. If NACCO will never be able to utilize the tax deductions associated with asset ownership, T = 0, and Equation (25.2) becomes:

$$\text{NAL} = 10M - \sum_{t=1}^{10} \frac{1.7M}{[1+0.116]^t} - \frac{0.5M}{[1+0.17]^{10}} = \$131{,}260$$

NACCO should lease in this case also.

A10. a. The equivalent-loan approach to the lease-versus-borrow-and-buy decision involves comparing the amount of financing provided by the lease to the amount of financing provided by the equivalent loan, where the equivalent loan is the maximum amount the lessee could borrow if it dedicates the future incremental cash flow stream to service conventional secured debt.

b. The equivalent loan is the present value of the net cash flows at the firm's after-tax cost of debt. Using Equation (25.4):

$$\text{EL} = \frac{10M}{(1.066)} + \sum_{t=2}^{5} \frac{7M}{(1.066)^t} = 31.83M$$

A11. a. The after-tax cost of debt is $(1 - 0.34)\times 12\% = 7.92\%$, and the net advantage to leasing is:

$$\text{NAL} = 1M - \frac{0.3M}{(1.0792)} - \frac{0.275M}{(1.0792)^2} - \frac{0.250M}{(1.0792)^3} - \frac{0.225M}{(1.0792)^4} - \frac{0.2M}{(1.0792)^5} = -\$15{,}496$$

b. The internal rate of return for the lease is the rate that makes the net advantage equal to zero:

$$0 = 1M - \frac{0.3M}{(1+\text{IRR})} - \frac{0.275M}{(1+\text{IRR})^2} - \frac{0.250M}{(1+\text{IRR})^3} - \frac{0.225M}{(1+\text{IRR})^4} - \frac{0.2M}{(1+\text{IRR})^5}$$

IRR = 8.55%

c. The equivalent loan is:

$$\text{EL} = \frac{0.3M}{(1.0792)} + \frac{0.275M}{(1.0792)^2} + \frac{0.250M}{(1.0792)^3} + \frac{0.225M}{(1.0792)^4} + \frac{0.2M}{(1.0792)^5} = \$1{,}015{,}496$$

d. Allied should borrow and buy the asset.

A12. **Project financing** involves financing a large capital investment project on a stand-alone basis. Project financing is an appropriate method of financing when (1) the project consists of a discrete asset or a discrete set of assets that are capable of standing alone as an independent economic unit, and (2) the economic prospects of the project, combined with commitments from the sponsors and/or from third parties, assure that the project will generate sufficient revenue net of operating costs to service project debt.

A13. The principal advantages of project financing are (1) risk sharing, (2) expanded debt capacity, and (3) a lower cost of debt. The principal disadvantages of project financing are the additional costs of arranging such financing--transaction costs, a yield premium for the higher agency costs, and a yield premium to compensate for any business risk the lenders must assume.

A14. Lease financing and limited partnership financing are both tax-oriented financing methods that provide a cost-effective alternative to conventional financing when the firm is unable to utilize fully the tax deductions and tax credits of asset ownership.

A15. We modify the analysis in Table 25-4 so that the limited partners receive 90% of partnership income, losses, tax credits, and cash distributions for the life of the limited partnership. Columns (1) - (3) are unchanged, and the new values for columns (4) - (7) are presented below:

End of Year	Distribution to Limited Partners	Tax on Income (Loss) Forgone	Residual Value Forgone	Net Cash Flow to Sponsor
0	----	----	----	$45.00
1	$1.35	-$1.26	----	-2.61
2	2.70	-0.72	----	-3.42
3	3.15	-0.54	----	-3.69
4	6.30	0.72	----	-5.58
5	9.00	1.80	----	-7.20
6	10.80	2.52	----	-8.28
7	12.60	3.24	----	-9.36
8	14.40	3.96	----	-10.44
9	15.75	4.50	----	-11.25
10	17.10	5.04	$92.34	-104.40

Cost of limited partner capital = 17.71%.

B1. The acquisition of an asset is analyzed by computing a conventional NPV using the lessee's weighted average cost of capital as the discount rate. The choice of leasing versus borrowing money to buy the asset is analyzed by computing the net advantage to leasing. And, because the contractual lease payments are similar in risk to promised debt payments, the appropriate discount rate for this decision is the lessee's cost of secured debt. Finally, the total NPV of leasing to acquire the asset is the conventional NPV plus the net advantage to leasing.

B2. To evaluate the net advantage to the lessor of entering into the lease, the lessor must use the discount rate that is appropriate for the riskiness of the cash flows. It is the lessee's cost of secured debt (as determined in the capital market) that measures the riskiness of the lease payments so the lessor should discount the after-tax net cash flows at $(1 - T)$ times the lessee's cost of secured debt, where T is the lessor's marginal tax rate.

B3. a. Assuming zero residual value, annual depreciation is $100,000 (= 1M/10). Using Equation (25.2):

$$NAL = 1M - \sum_{t=1}^{10} \frac{(1-0.35)\times(135{,}000) + 0.35\times(100{,}000)}{[1+(1-0.35)\times(0.10)]^t} = \$117{,}571$$

The firm should lease.

b. Assuming a $250,000 residual value, annual depreciation is $75,000 (= [1M - 0.25M]/10.

$$NAL = 1M - \sum_{t=1}^{10} \frac{(1-0.35)\times(135{,}000) + 0.35\times(75{,}000)}{[1+(1-0.35)\times(0.10)]^t} - \frac{250{,}000}{(1+0.15)^{10}} = \$118{,}678$$

The firm should still lease.

B4. a. Using Equation (25.3) and solving for the IRR:

$$NAL = 1M - \sum_{t=1}^{10} \frac{(1-0.35)\times(135{,}000) + 0.35\times(75{,}000)}{[1 + IRR]^t} - \frac{250{,}000}{(1+IRR)^{10}} = 0$$

IRR = 5.6%

b. The IRR of the lease is that discount rate which sets the NAL = 0. The IRR of 5.6% is less than the firm's after tax cost of secured debt of 6.5%, so the firm should lease.

c. Using Equation (25.4):

$$EL = \sum_{t=1}^{10} \frac{(1-0.35)\times(135{,}000) + 0.35\times(75{,}000)}{[1+0.065]^t} - \frac{250{,}000}{(1+0.065^{10}} = \$881{,}322$$

The equivalent loan approach gives the amount the lessee could borrow if it dedicated all of its after-tax lease payments to servicing the borrowing.

B5 Using Equation (25.2) and assuming the computer has zero salvage value at the end of the 5 years:

$$NAL = 5000 - \sum_{t=0} \frac{100}{(1+0.08/12)^t} = \$102.85$$

You should gracefully decline your mother's offer and lease the PC.

B6. a. The annual depreciation on the trolley is $2500 (= [25,000 - 5000]/8) and the resulting depreciation tax credit is $1000 (= 0.4 ×[2500]). The after-tax lease payments are $3000 (= (1 - 0.4)×5000), so the net after-tax cash outflows for years 1-8 (after-tax lease payments plus forgone depreciation) are $4000 with an additional outflow of $5000 in year 8 representing the forgone net salvage value.

Year	Net Cash Flow to Lessee
0	+25,000
1-7	-4000
8	-9000

b. The net advantage to leasing is computed by applying Equation (25.2):

$$\text{NAL} = 25{,}000 - \sum_{t=1}^{8} \frac{4000}{[1+(1-0.4)\times 0.12]^t} - \frac{5000}{(1.16)^8} = -\$226.30$$

Lake Trolley should borrow and buy because NAL < 0.

c. With a tax rate of zero, the net cash outflows are \$5000 for years 1-8 with an additional \$5000 outflow in year 8. The net advantage to leasing is:

$$\text{NAL} = 25{,}000 - \sum_{t=1}^{8} \frac{5000}{(1+0.12)^t} - \frac{5000}{(1.20)^8} = -\$1001.04$$

Lake Trolley should borrow and buy because NAL < 0.

d. For part b, the net salvage value would be \$2300 (= (1 - 0.4)×500 + 0.4 ×(5000)), and NAL would be:

$$\text{NAL} = 25{,}000 - \sum_{t=1}^{8} \frac{4000}{[1+(1-0.4)\times 0.12]^t} - \frac{2300}{(1.16)^8} = \$597.24$$

For part c, the net salvage value would be \$500, and the NAL would be:

$$\text{NAL} = 25{,}000 - \sum_{t=1}^{8} \frac{5000}{(1+0.12)^t} - \frac{500}{(1.20)^8} = \$45.52$$

Now, Lake Trolley should lease the asset because, in either case, the NAL > 0.

B7. If the lease payments must be made at the beginning of each year, then the payments will occur at T = 0 through $T = N - 1$. Equation (25.2) becomes:

$$\text{NAL} = P - \sum_{t=0}^{N-1} \frac{(1-T)\times(L_t - \Delta E_t) + T\times D_t}{[1+(1-T)\times r^*]^t} - \frac{\text{SAL}}{[1+r]^N} - \text{ITC}$$

Equation (25.3) becomes:

$$0 = P - \sum_{t=0}^{N-1} \frac{(1-T)\times(L_t - \Delta E_t) + T\times D_t}{[1 + \text{IRR}]^t} - \frac{\text{SAL}}{[1 + \text{IRR}]^N} - \text{ITC}$$

and Equation (25.4) becomes:

$$\text{EL} = \sum_{t=0}^{N-1} \frac{(1-T)\times(L_t - \Delta E_t) + T\times D_t}{[1+(1-T)\times r^*]^t} + \frac{\text{SAL}}{[1+r]^N}$$

B8. a. Using the modified Equation (25.2) from the solution to B7, we can compute the net advantage to leasing:

$$\text{NAL} = P - \sum_{t=0}^{N-1} \frac{(1-T)\times(L_t - \Delta E_t) + T\times D_t}{[1+(1-T)\times r^{\prime}]^t} - \frac{\text{SAL}}{[1+r]^N} - \text{ITC}$$

$$= 35{,}000 - \sum_{t=0}^{4} \frac{(1-0.34)\times(7850-0)+0.34\times(6000)}{[1+(1-0.34)\times 0.10]^t} - \frac{5000}{[1.12]^5} = \$260.31$$

Because NAL > 0, New Horizon should lease the truck.

b. With $T = 0$, the net advantage to leasing is:

$$NAL = 35{,}000 - \sum_{t=0}^{4} \frac{7850}{[1.10]^t} - \frac{5000}{[1.16]^5} = -\$114.01$$

Because NAL < 0, New Horizon should borrow and buy the truck.

c. The NAL calculation is situation-specific, and the timing of lease payments and tax realizations can alter the usual logic that leasing is more advantageous to a lessee who cannot utilize the tax benefits of ownership.

B9. a. With lease payments and tax credits at the beginning of each year, the NAL is:

$$\text{NAL} = 100{,}000 - \sum_{t=0}^{3} \frac{22{,}000}{[1.14]^t} - \frac{20{,}000}{[1.20]^4} = \$17{,}279$$

b. The amount of the equivalent loan is:

$$\text{EL} = \sum_{t=0}^{3} \frac{22{,}000}{[1.14]^t} + \frac{20{,}000}{[1.20]^4} = \$82{,}721$$

c. The internal rate of return for the lease is the discount rate, IRR, that solves the following equation:

$$0 = 100{,}000 - \sum_{t=0}^{3} \frac{22{,}000}{[1 + \text{IRR}]^t} - \frac{20{,}000}{[1 + \text{IRR}]^4}$$

IRR = 4.08%

B10. Using Equation (25.2), if the machine's required return is 16%, the net advantage of leasing is:

$$\text{NAL} = 800{,}000 - \sum_{t=1}^{10} \frac{(1-0.4)\times 110{,}000 + 0.4\times(75{,}000)}{[1+(1-0.4)\times 0.12]^t} - \frac{50{,}000}{[1.16]^{10}} = \$120{,}592$$

Because the NAL > 0, leasing is preferable to borrowing and buying.

B11. a. If Empire buys the trucks, the annual depreciation will be \$150,000 (= 750,000/5) and the resulting depreciation tax credit is \$51,000 (=0.34×(150,000)). The after-tax lease payments are \$125,400 (= (1 - 0.34)×190,000), so the annual after-tax cash outflow for years 1 through

5 will be $176,400. The forgone salvage value creates an additional year 5 outflow of $100,000. The stream of net cash flows to Empire under the lease financing is summarized below:

Time	CFAT
0	$750,000
1-4	-176,400
5	-276,400

b. The net advantage to leasing is:

$$NAL = 750{,}000 - \sum_{t=1}^{5} \frac{176{,}400}{[1+(1-0.34)\times 0.14]^t} - \frac{100{,}000}{[1.16]^5} = \$20{,}506$$

c. The amount of the equivalent loan is:

$$EL = \sum_{t=1}^{5} \frac{176{,}400}{[1+(1-0.34)\times 0.14]^t} + \frac{100{,}000}{[1.16]^5} = \$729{,}494$$

d. Empire should lease the fleet of trucks because the net advantage to leasing is positive and because the amount of the equivalent loan that could be supported by the stream of lease cash flows is less than the cost of the trucks.

B12. a. The net advantage to leasing is:

$$NAL = 10.5 - \frac{3.0}{(1.13)} - \frac{5.0}{(1.13)^2} - \frac{5.0}{(1.13)^3} = \$0.46 \text{ million}$$

The NAL is positive, so leasing is advantageous.

b. The amount of the equivalent loan is:

$$EL = \frac{3.0}{(1.13)} + \frac{5.0}{(1.13)^2} + \frac{5.0}{(1.13)^3} = \$10.04 \text{ million}$$

c. The internal rate of return is the discount rate, IRR, that makes the net advantage to leasing zero:

$$0 = 10.5 - \frac{3.0}{(1 + IRR)} - \frac{5.0}{(1 + IRR)^2} - \frac{5.0}{(1 + IRR)^3}$$

$$IRR = 10.58\%$$

d. The three approaches yield consistent results. The equivalent loan is simply a mathematical rearrangement of the net advantage to leasing (Cost - EL = NAL) so it must yield the same results. The IRR is the discount rate that causes the NAL to equal zero, but it ignores the difference in the discount rate for the salvage value. If there is only one change in sign in the net cash flow stream and if the salvage value is zero, the IRR method will give results identical to the other two methods.

B13. a. The lease payments and depreciation tax credits, and resulting net cash flows (excluding the residual value) are summarized in the table below:

Year	L_t	$(1-T)\times L_t$	D_t	$T\times D_t$	$CFAT_t$
1	1000	600	2000	800	1400
2	1000	600	1750	700	1300
3	1000	600	1500	600	1200
4	1000	600	1250	500	1100
5	1000	600	1000	400	1000
6	2000	1200	400	160	1360
7	2000	1200	400	160	1360
8	2000	1200	400	160	1360
9	2000	1200	400	160	1360
10	2000	1200	400	160	1360

$$NAL = 10{,}000{,}000 - \sum_{t=1}^{10} \frac{CFAT_t}{[1+(1-0.4)\times 0.125]^t} - \frac{500{,}000}{[1.15]^{10}}$$

$$= 10{,}000{,}000 - 8{,}746{,}197 - 123{,}592 = \$1{,}130{,}211$$

b. If the firm is unable to claim the tax credits in years 1 through 3, it loses the opportunity to claim tax deductions of $400 per year on the $1,000 lease payments and also loses the depreciation tax credits of $800, $700, and $600 in years 1 through 3, respectively. The total tax loss carryforward is $3300 (= 3 ×(400) + 800 + 700 + 600), which will be claimed in year 4. The net cash flows (excluding the residual value) will be as follows:

Year	L_t	$(1-T)\times L_t$	D_t	$T\times D_t$	$CFAT_t$
1	1000	1000	2000	0	1000
2	1000	1000	1750	0	1000
3	1000	1000	1500	0	1000
4	1000	600	1250	500	-2200
5	1000	600	1000	400	1000
6	2000	1200	400	160	1360
7	2000	1200	400	160	1360
8	2000	1200	400	160	1360
9	2000	1200	400	160	1360
10	2000	1200	400	160	1360

The net advantage to leasing is:

$$NAL = 10{,}000{,}000 - \sum_{t=1}^{10} \frac{CFAT_t}{[1+(1-0.4)\times 0.125]^t} - \frac{500{,}000}{[1.15]^{10}}$$

$$= \$10{,}000{,}000 - 5{,}482{,}470 - 123{,}592 = \$4{,}393{,}938$$

B14. a. Amalgamated's break-even lease rate is the lease rate that makes the NPVL, Equation (25.5), zero. First, we must compute the appropriate cost of debt. With a 120% collateralized loan, Sandoval would be able to borrow $4,166,667 (= 5,000,000/1.20) at the secured rate and the remaining $833,333 would be borrowed at the unsecured debt rate.

$$r' = (4{,}166{,}667/5{,}000{,}000)10\% + (833{,}333/5{,}000{,}000)12\% = 10.33\%$$

The annual depreciation will be \$791,667 (= (5,000,000-250,000)/6). The break-even lease rate is the payment, L_R, that solves the following equation:

$$0 = -5{,}000{,}000 + \sum_{t=1}^{6} \frac{(1-0.4)\times L_R + 0.4\times(791{,}667)}{[1+(1-0.4)\times 0.1033]^t} + \frac{250{,}000}{[1.15]^6}$$

so that $L_R = \$1{,}140{,}606$

b. Sandoval's net advantage to leasing at Amalgamated's break-even lease rate is:

$$\text{NAL} = 5{,}000{,}000 - \sum_{t=1}^{6} \frac{(1-0.3)\times 1{,}140{,}606 + 0.3\times(791{,}667)}{[1+(1-0.3)\times 0.1033]^t} - \frac{250{,}000}{[1.15]^6} = -\$10{,}840$$

c. Sandoval's break-even lease rate is the payment, L_E, that makes the net advantage to leasing zero:

$$0 = 5{,}000{,}000 - \sum_{t=1}^{6} \frac{(1-0.3)\times L_E + 0.3\times(791{,}667)}{[1+(1-0.3)\times 0.1033]^t} - \frac{250{,}000}{[1.15]^6}$$

so that $L_E = \$1{,}137{,}334$

d. Because Amalgamated's break-even lease rate exceeds Sandoval's break-even lease rate, it is not possible for Amalgamated and Sandoval to find a mutually beneficial lease rate.

B15. a. The mining firm is acting as agent for the lender, while having control over the separate corporation. Situations could arise in which the goal of the mining firm to maximize its shareholders' wealth could diverge from the goal of the lender to be repaid. For example, although the copper purchase agreement requires that the mining firm purchase the output of the separate corporation, the mining firm may be able to manipulate the amount of output or price paid for the output to its advantage. At the extreme, the mining firm could shut down the project and allow the separate corporation to default on its debt. Since the loan is nonrecourse, the mining firm is not responsible for the losses to the bank. Goal divergence such as this leads to additional contracting and monitoring costs as well.

b. The agency costs mentioned above are not present when the firm itself promises to pay the loan. The bank would expect to be compensated for these costs in a loan to the separate corporation.

B16. The hell-or-high water contract requires that the triple-A rated firm pay for the copper output under all circumstances, whether the copper is available for delivery or not. Thus, the cash flows from the mine are actually cash flows promised by the triple-A firm. A lender would recognize this and lend to the mining firm at a rate closer to that of the triple-A firm.

B17. a. Apply Equation (25.7) and solve by trial-and-error for CLPC = 9.86%.

$$\text{NP} = \frac{\text{CFAT}_t}{(1 + \text{CLPC})^t}$$

Or, on your calculator, solve for IRR = 9.86%.

b. Yes, the firm should employ the limited partnership financing, since it is cheaper at 9.86% than conventional equity financing at 15%.

B18. a. The cost of limited partnership capital when the partnership has debt is analogous to the cost of leveraged equity.

b. If a limited partnership has debt, its cost of limited partner capital should be compared to the required return on a conventionally financed and identically leveraged investment.

B19. The net proceeds of the investment represent a $5.0 million cash inflow to the general partner at time zero. The residual value is $21 million (= 7 ×[3]) and, because the assets will have been fully depreciated at that time, the residual value net of taxes will be $12.6 million. Seventy-five percent of that ($9.45 million) will be distributed to the limited partners at the end of year 8. The other cash flows associated with the partnership are summarized below (cash amounts in millions):

End of Year	Partnership Taxable Income	Distribution to Limited Partners	Tax on Income (Loss) Forgone	Residual Value Forgone	Net Cash Flow to Sponsor
0	----	----	----	----	$5.0000
1	-$0.375	$0.1875	-$0.1125	----	-0.3000
2	0.375	0.7500	0.1125	----	-0.6375
3	0.875	1.1250	0.2625	----	-0.8625
4	1.375	1.5000	0.4125	----	-1.0875
5	1.625	1.6875	0.4875	----	-1.2000
6	1.875	1.8750	0.5625	----	-1.3125
7	2.125	2.0625	0.6375	----	-1.4250
8	2.375	2.2500	0.7125	9.45	-10.9875

The partnership taxable income is the operating cash flow less the straight-line depreciation of $625,000 (= 5,000,000/8); the distribution to limited partners is 75% of the operating cash flow; and the tax on income (loss) forgone is the tax rate applied to 75% of the partnership taxable income. The net cash flow to sponsor is the tax on income (loss) forgone less distribution to limited partners and includes the net proceeds in year 0 and the $9.45 million of the residual value forgone in year 8. Solving for the internal rate of return of the net cash flow stream gives the cost of limited partner capital of 22.31%.

B20. a. Using Equation (25.2) with $T = 0$:

$$\text{NAL} = 20{,}000 - 2{,}000 - \sum_{t=1}^{48} \frac{360}{[1+0.01]^t} - \frac{6000}{[1.01]^{48}} = \$471$$

The NAL is positive, so lease the car.

b. Using Equation (25.5) with monthly depreciation of $292.67 (= [20,000 - 6000]/48):

$$\text{NPVL} = -20{,}000 + 2{,}000 + \sum_{t=1}^{48} \frac{360\times(1-0.4) + 0.4\times 292.67}{[1+(1-0.4)\times 0.01]^t} + \frac{6000}{[1.01]^{48}} = \$1526.45$$

Yes, Honda benefits from the lease too.

c. When it acts as a lessor, Honda can claim depreciation and other tax credits which are unavailable to it when it sells the car outright. Because the leasing market is competitive, Honda passes some of these savings along to lessees.

C1. a. Lake Trolley Company's **break-even tax rate** is the rate, T, that solves the following equation:

$$0 = 25{,}000 - \sum_{t=1}^{8} \frac{(1-T)\times 5000 + T\times 3062.5}{[1+(1-T)\times 0.12]^t} - \frac{500}{(1.16)^8}$$

By trial and error, we find $T = 1.2\%$

b. Lake Trolley Company's effective tax rate, rather than the statutory tax rate, is the relevant tax rate for this analysis. With a break-even tax rate of 1.2%, leasing on these terms would be advantageous for Lake Trolley Company only if its average effective tax rate over the term of the lease is essentially zero.

C2. a. A lease represents a form of secured debt. Each lease payment includes an interest component and a principal repayment component. Because the lease payment is taxable as ordinary income to the lessor, in effect, the entire debt service stream (i.e., the interest component and the principal component) is taxable as ordinary income.

b. The following condition must hold:

PV(lessee's lease payment tax shields) + PV(lessor's depreciation tax shields + other tax credits)

is greater than

PV(lessor's lease payment tax liability) + PV(lessee's depreciation tax shields + other tax credits) in order for a financial lease transaction to generate positive net present value benefits for lessor and lessee combined.

c. Each after-tax lease payment is \$1,047,000 (= 1,745,000 - 698,000). The net advantage to leasing for NACCO must decrease because the lease payments are accelerated by 1 year:

$$NAL = 10{,}000{,}000 - \sum_{t=0} \frac{1{,}047{,}000}{[1+(1-0.4)\times 0.12]^t} - \sum_{t=1} \frac{380{,}000}{[1+(1-0.4)\times 0.12]^t} - \frac{500{,}000}{[1+0.15]^{10}}$$

$$= 10{,}000{,}000 - 7{,}810{,}789 - 2{,}644{,}460 - 123{,}592 = -\$578{,}841$$

which is more than \$500,000 lower than the net advantage to leasing calculated in the text, which was -\$54,236.

d. If NACCO is nontaxable, its net advantage to leasing is:

$$NAL = 10{,}000{,}000 - \sum_{t=0}^{9} \frac{1{,}745{,}000}{(1.12)^t} - \frac{500{,}000}{(1.175)^{10}} = -\$1{,}142{,}472$$

The net advantage of the lease to the lessor is:

$$NPVL = -10{,}000{,}000 + \sum_{t=0}^{9} \frac{1{,}047{,}000}{(1.072)^t} + \sum_{t=1}^{10} \frac{380{,}000}{(1.072)^t} + \frac{500{,}000}{(1.15)^{10}} = \$578{,}841$$

which is smaller than NAL in absolute value.

e. A mutually advantageous lease rate, L, must satisfy the following two conditions simultaneously:

$$\text{NAL} = 10{,}000{,}000 - \sum_{t=0}^{9} \frac{L}{(1.12)^t} - \frac{500{,}000}{(1.175)^{10}} > 0 \rightarrow L < 1{,}564{,}465$$

$$\text{NPVL} = -10{,}000{,}000 + \sum_{t=0}^{9} \frac{0.6 \times L}{(1.072)^t} + \sum_{t=1}^{10} \frac{380{,}000}{(1.072)^t} + \frac{500{,}000}{(1.15)^t} > 0 \rightarrow L > 1{,}615,$$

The two conditions cannot be satisfied simultaneously; a mutually beneficial lease rate therefore does not exist.

f. Making the lease payments in advance imposes a net tax cost when the lessor pays income tax at a higher rate than the lessee.

C3. The debt service parity principle requires computing the outflows associated with the lease and then "selling" these outflows to determine the amount of debt that could be supported by that stream of debt service payments.

Time	Item	CFBT	CFAT
1-N	Lease payments	L_t/yr	$(1-T) \times L_t$
1-N	Increased expenses	ΔE_t/yr	$(1-T) \times \Delta E_t$
1-N	Depreciation	0/yr	$T \times D_t$
N	Salvage value	SAL	SAL

The expected salvage value could be "sold" at a discount rate equal to the required return on the project, while the other cash flows could be "sold" at the after-tax costs of the loan that is being made. The total present value is then given by Equation (25.4).

C4. a. The net present value of the lease to the lessor is given by

$$\text{NPV}_R = -C + (1-T_R) \times L + \frac{(1-T_R) \times L}{1 + (1-T_R) \times D} + \frac{T_R \times C/2}{1 + (1-T_R) \times D} + \frac{T_R \times C/2}{[1 + (1-T_R) \times D]^2}$$

$$= -C + \frac{(1-T_R) \times L[1 + (1-T_R) \times D] + T_R \times C/2}{1 + (1-T_R) \times D} + \frac{(1-T_R) \times L \times [1 + (1-T_R) \times D] + T_R \times C/2}{[1 + (1-T_R) \times D]^2}$$

$$= -C + \left[\frac{(1-T_R) \times L \times [1 + (1-T_R) \times D] + T_R \times C/2}{(1-T_R) \times D}\right]\left[1 - \left[\frac{1}{1 + (1-T_R) \times D}\right]^2\right]$$

b. The net present value of the lease to the lessee is given by

$$\text{NPV}_E = C - (1-T_E) \times L - \frac{(1-T_E) \times L}{1 + (1-T_E) \times D} - \frac{T_E \times C/2}{1 + (1-T_E) \times D} - \frac{T_E \times C/2}{[1 + (1-T_E) \times D]^2}$$

$$= C - \left[\frac{(1-T_E) \times L \times [1 + (1-T_E) \times D] + T_E \times C/2}{(1-T_E) \times D}\right]\left[1 - \left[\frac{1}{1 + (1-T_E) \times D}\right]^2\right]$$

c. If $T_E = T_R$, then the expression for NPV_E in part b becomes

$$NPV_E = C - \left[\frac{(1-T_R)\times L\times[1+(1-T_R)D]+T_R\times C/2}{(1-T_R)\times D}\right]\left[1-\left[\frac{1}{1+(1-T_R)\times D}\right]^2\right]$$

and $NPV_E + NPV_R = 0$ so that leasing is a zero-sum game. There might nevertheless be other reasons to lease, for example, a lessee's desire to avoid the risk of technological obsolescence, which an equipment manufacturer, serving as lessor, might be willing to bear at a cost that is less than the lessee's cost of bearing such risk.

d. If $T_E = 0 < T_R$, then the net present value of the lease to the lessee becomes

$$NPV_E = C - \left[\frac{L\times(1+D)}{D}\right]\times\left[1-\left[\frac{1}{(1+D)}\right]^2\right]$$

Define L_E as the solution to the equation $NPV_E = 0$, and L_R as the solution to the equation $NPV_R = 0$. If $T_R = 0$, then

$$NPV_R = -C + \left[\frac{L\times(1+D)}{D}\right]\times\left[1-\left[\frac{1}{(1+D)}\right]^2\right] = -NPV_E$$

It follows that $L_E = L_R$

If we can show that $dL_R/dT_R > 0$, then it will be impossible to find a lease rate L for which $NPV_R > 0$ simultaneously because $L_R > L_E$ would hold for all $T_R > 0$.

While we do not offer a formal mathematical proof, you can easily satisfy yourself that $dL_R/dT_R > 0$ by testing various values of D and T_R. Consider for example:

	D=0.02		D=0.10	
T_R	L_E	L_R	L_E	L_R
0	0.504950*C*	0.504950*C*	0.523810*C*	0.523810*C*
0.05	0.504950*C*	0.505444*C*	0.523810*C*	0.526149*C*
0.10	0.504950*C*	0.505938*C*	0.523810*C*	0.528511*C*
0.15	0.504950*C*	0.506433*C*	0.523810*C*	0.530893*C*
0.20	0.504950*C*	0.506929*C*	0.523810*C*	0.533298*C*
.				
.				
.				
0.40	0.504950*C*	0.508923*C*	0.523810*C*	0.543140*C*

Since $dL_R/dT_R > 0$, we conclude that $L_R > L_E$ for all $T_R > 0$ and that it is not possible to find an annual lease payment amount L for which $NPV_R > 0$ and $NPV_E > 0$ hold simultaneously.

e. Recalculating NPV_R and NPV_E when lease payments are made in arrears gives:

$$NPV_R = -C + \frac{(1-T_R)\times L + T_R\times C/2}{1+(1-T_R)\times D} + \frac{(1-T_R)\times L + T_R\times C/2}{[1+(1-T_R)\times D]^2}$$

$$= -C + \left[\frac{(1-T_R)\times L + T_R\times C/2}{(1-T_R)\times D}\right]\times\left[1 - \left[\frac{1}{1+(1-T_R)\times D}\right]^2\right]$$

$$NPV_E = C - \frac{(1-T_E)\times L + T_E\times C/2}{1+(1-T_E)\times D} + \frac{(1-T_E)\times L + T_E\times C/2}{[1+(1-T_E)\times D]^2}$$

$$= C - \left[\frac{(1-T_E)\times L + T_E\times C/2}{(1-T_E)\times D}\right]\times\left[1 - \left[\frac{1}{1+(1-T_E)\times D}\right]^2\right]$$

Suppose $T_R = 0.4$ and $T_E = 0$. To construct a numerical example in which $NPV_R > 0$ and $NPV_E > 0$ hold simultaneously, assume $D = 0.10$.

Given these assumptions, we have

$$NPV_E = C - \frac{L_E}{D}\times\left[1 - \left[\frac{1}{1+D}\right]^2\right] = 0$$

implies that

$$L_E = \frac{CD}{1 - \left[\frac{1}{1+D}\right]^2} = 0.57619C$$

If we can show that $L_R < L_E$ under the same assumptions, then any lease rate L such that $L_R < L < L_E$ will enable $NPV_R > 0$ and $NPV_E > 0$ to hold simultaneously.

Setting $NPV_R = 0$ and solving for L_R gives

$$L_R = \frac{C\times D}{1 - \left[\frac{1}{1+(1-T_R)\times D}\right]^2} - \frac{T_R\times C/2}{1-T_R}$$

Substituting $T_R = 0.4$ and $D = 0.10$ gives $L_R = 0.57573C$. Any lease payment amount L such that $0.57573C < L < 0.57619C$ will be mutually beneficial to lessor and lessee.

C5. a. If the bank's ratio of capital to assets is 8%, the bank cannot increase its assets unless it first increases its capital. A capitalized lease obligation gives rise to both an asset and a long-term liability. Thus, if the bank enters into a capital lease, each dollar of capitalized lease obligations gives rise to one dollar of (leased) assets, which must displace one dollar of earning assets (or other assets) in order to keep the bank's ratio of capital-to-assets at 8%.

b. If the bank is capital-constrained, an operating lease avoids the displacement of debt (e.g., deposits) and earning assets that would occur under a capital lease or if the asset were financed on a conventional basis. A bank that takes in a dollar of deposits and invests the dollar in earning assets will realize an interest margin m equal to the difference between its marginal return on assets and its marginal cost of funds. Equation (25.2) should be modified to reflect this factor by adding the term

$$\sum_{t=1}^{N} \frac{(1 - T)\times m\times A_t}{(1 + r)^t}$$

where N, r, and T are as defined in Equation (25.2) and A_t denotes the amount of earning assets (and deposits) that a capital lease or a conventional financing would displace during period t. This adjustment is interpreted in the following manner: The operating lease avoids the displacement of an amount A_t of deposits and earning assets during each period t. Having these assets would enable the bank to earn a pre-tax interest margin of $m\times A_t$ and an after-tax interest margin of $(1 - T)\times m\times A_t$ during period t. The period-by-period after-tax amounts are discounted at the bank's cost of capital because the riskiness of the incremental after-tax net interest reflects the overall riskiness of the bank.

c.

$$\text{NAL} = 736{,}000 - \left[\sum_{t=1}^{30} \frac{13{,}600}{(1+0.13/12)^t} + \sum_{t=31}^{60} \frac{16{,}600}{(1+0.13/12)^t}\right] - \frac{73{,}600}{(1+0.09)^{10}} + \sum_{t=1}^{10} \frac{(0.02\times A_t)/2}{(1+0.09)^t}$$

$$= 51{,}827 + \frac{0.01\times(736{,}000)}{(1.09)} + \frac{0.01\times(669{,}760)}{(1.09)^2} + \frac{0.01\times(603{,}520)}{(1.09)^3} + \frac{0.01\times(537{,}280)}{(1.09)^4} + \frac{0.01\times(471{,}0}{(1.09)^5}$$

$$+ \frac{0.01\times(404{,}800)}{(1.09)^6} + \frac{0.01\times(338{,}560)}{(1.09)^7} + \frac{0.01\times(272{,}320)}{(1.09)^8} + \frac{0.01\times(206{,}080)}{(1.09)^9} + \frac{0.01\times(139{,}840)}{(1.09)^{10}}$$

$$= 51{,}827 + 31{,}089 = \$82{,}916$$

d. NAL = \$51,827, since there is no capital-constraint-avoidance benefit.

CHAPTER 26 -- DERIVATIVES AND HEDGING

Note: The notation here uses T in place of Δt in the option pricing formulas.

A1. A convertible bond provides its owner the rights of a bondholder along with the right to convert the bond into shares of the issuer's common stock at a stated **conversion price**. This conversion price is analogous to a warrant's exercise price. The market value of a convertible bond equals the sum of its straight bond value and the value of the conversion option. The value of the conversion option is essentially the same as the value of warrants on the number of shares into which the bond can be converted at the conversion price. The owner of a convertible bond thus can be viewed as owning a straight bond and warrants on the firm's stock.

A2. We apply the Black-Scholes OPM, Equations (26.2)-(26.4), with $P_0 = \$28$, $\sigma = 0.40$, APR = 0.06, $S = 30$, and $T = 3/4$:

$$d_1 = \frac{\ln(P_0/S)}{\sigma \times T^{1/2}} + \frac{T \times \text{APR}}{\sigma \times T^{1/2}} + \frac{\sigma \times T^{1/2}}{2} = \frac{\ln(28/30)}{0.40 \times (3/4)^{1/2}} + \frac{(3/4) \times (0.06)}{0.40 \times (3/4)^{1/2}} + \frac{0.40 \times (3/4)^{1/2}}{2} = 0.1039$$

$$d_2 = d_1 - \sigma \times T^{1/2} = 0.1039 - 0.40 \times (3/4)^{1/2} = -0.2425$$

From Appendix B, $N(d_1) = N(0.1039) = 0.5413$ and $N(d_2) = N(-0.2425) = 0.4042$. Therefore,

$$\text{CALL} = P_0 \times N(d_1) - S \times N(d_2) \times e^{-T \times \text{APR}} = 28 \times (0.5411) - 30 \times (0.4042) \times e^{-(0.75) \times (0.06)} = \$3.56$$

A3. We apply the Black-Scholes OPM with $P_0 = \$28$, $\sigma = 0.40$, APR = 0.06, $S = 30$, and $T = 2$:

$$d_1 = \frac{\ln(P_0/S)}{\sigma \times T^{1/2}} + \frac{T \times \text{APR}}{\sigma \times T^{1/2}} + \frac{\sigma \times T^{1/2}}{2} = \frac{\ln(28/30)}{0.40 \times (2)^{1/2}} + \frac{2 \times (0.06)}{0.40 \times (2)^{1/2}} + \frac{0.40 \times (2)^{1/2}}{2} = 0.3730$$

$$d_2 = d_1 - \sigma \times T^{1/2} = 0.3730 - 0.40 \times (2)^{1/2} = -0.1927$$

From Appendix B, $N(d_1) = N(0.3730) = 0.6454$ and $N(d_2) = N(-0.1927) = 0.4236$. Therefore,

$$\text{CALL} = P_0 \times N(d_1) - S \times N(d_2) \times e^{-T \times \text{APR}} = 28 \times (0.6454) - 30 \times (0.4236) \times e^{-(2) \times (0.06)} = \$6.80$$

A4. A **derivative** is a financial instrument whose value depends upon the value of some other asset or financial instrument.

A5. A **convertible** security is a debt or preferred stock instrument that can be exchanged, at the owner's option, for a stated number of shares of the issuing company's common stock.

A6. An **interest rate swap** is a transaction in which two companies swap interest payment obligations. The objective of both parties is to lower their interest costs. In a typical interest rate swap arrangement, two companies borrow equal amounts simultaneously: the stronger credit sells fixed-interest-rate bonds; the weaker credit borrows at a floating interest rate under a loan agreement of matching maturity; and they swap interest payment obligations with the stronger credit company agreeing to pay floating rate and the weaker credit company agreeing to pay fixed-rate.

A7. A **warrant** is a long-term call option issued by a firm.

A8. A warrant is a long-term call option. The firm is giving up the value of the warrants when it includes them as a "sweetener" to a bond issue because it is in a short position. Therefore, since options are valuable, it is not costless for firms to include warrants with a bond issue because the warrants provide the bondholders the option to share in the future good fortunes of the firm along with the shareholders. If the share price increases sufficiently, the bondholders will exercise their warrants thereby diluting the claims of the shareholders.

A9. A futures contract can be considered to be a sequence of forward contracts, each day the old forward contract expires and a new one is written. The basic form of a futures contract is identical to that of a standardized forward contract. However, gains and losses on a futures contract are realized daily, while on a forward contract they are realized only on the contract settlement date. In addition, futures contracts are exchange-traded, while forward contracts are traded over-the-counter.

A10. The primary reason for a firm to engage in hedging would be to reduce its sensitivity to changes in the price of a commodity, foreign exchange rate, or interest rate.

A11. a. A floating rate borrower can use an interest rate swap to hedge against changes in interest rates by entering a swap to pay fixed and receive floating.

b. A lender can purchase an option to buy a bond as a hedge against a decline in interest rates.

c. An interest rate future contract (that is, a contract with a fixed-income security as its underlying asset) can be purchased by a lender who wants to hedge against a decline in rates or sold by a borrower who wishes to hedge against an increase in rates.

B1. Using the Simple Option Pricing Model from Section 26-4,

$$\text{CALL} = \left[\frac{1}{(1.05)^2}\right]\times(0.36\times[0] + 0.48\times[0] + 0.16\times[260 - 180]) = \$11.61$$

B2. Again applying the Simple Option Pricing Model from Section 26-4,

$$\text{CALL} = \left[\frac{1}{(1.05)^2}\right]\times(0.16\times[0] + 0.48\times[160 - 140] + 0.36\times[260 - 140]) = \$47.89$$

B3. An equation for put options analogous to the Simple Option Pricing Model on page 850 is:

$$\text{PUT} = \left[\frac{1}{(1.05)^2}\right]\times(0.36\times[185 - 60] + 0.48\times[185 - 160] + 0.16\times[0]) = \$51.70$$

B4. a. We apply the Black-Scholes OPM, Equations (26.2)-(26.4), with $P_0 = \$48.50$, $\sigma = 0.24$, APR $= 0.05$, $S = 50$, and $T = 8/12$:

$$d_1 = \frac{\ln(P_0/S)}{\sigma\times T^{1/2}} + \frac{T\times\text{APR}}{\sigma\times T^{1/2}} + \frac{\sigma\times T^{1/2}}{2} = \frac{\ln(48.50/50)}{0.24\times(8/12)^{1/2}} + \frac{(8/12)\times(0.05)}{.24\times(8/12)^{1/2}} + \frac{0.24\times(8/12)^{1/2}}{2} = 0.112647$$

$$d_2 = d_1 - \sigma\times T^{1/2} = 0.112647 - 0.24\times(8/12)^{1/2} = -0.083312$$

From Appendix B, $N(d_1) = N(0.11) = 0.5438$ and $N(d_2) = N(-0.08) = 0.4681$. Therefore,

$$\text{CALL} = P_0\times N(d_1) - S\times N(d_2)\times e^{-T\times\text{APR}} = 48.50\times(0.5438) - 50\times(0.4681)\times e^{-(8/12)\times(0.05)} = \$3.74$$

b. The put is valued using Put-Call Parity and Equation (26.7):

$$\text{PUT} = \text{CALL} + S\times e^{-T\times \text{APR}} - P_0 = 3.74 + 50\times e^{-(8/12)\times(0.05)} - 48.50 = \$3.60$$

c. Repeating part a with $T = 4/12$ gives:

$$d_1 = \frac{\ln(48.50/50)}{0.24\times(4/12)^{1/2}} + \frac{(4/12)\times(0.05)}{0.24\times(4/12)^{1/2}} + \frac{0.24\times(4/12)^{1/2}}{2} = -0.030257$$

$$d_2 = -0.030257 - 0.24\times(4/12)^{1/2} = -0.168821$$

From Appendix B, $N(d_1) = N(-0.03) = 0.4880$ and $N(d_2) = N(-0.17) = 0.4325$. Therefore,

$$\text{CALL} = P_0\times N(d_1) - S\times N(d_2)\times e^{-T\times \text{APR}} = 48.50\times(0.4880) - 50\times(0.4325)\times e^{-(4/12)\times(0.05)} = \$2.40$$

d. Using Put-Call Parity,

$$\text{PUT} = \text{CALL} + S\times e^{-T\times \text{APR}} - P_0 = 2.40 + 50\times e^{-(4/12)\times(0.05)} - 48.50 = \$3.07$$

B5. Using the Black-Scholes OPM with $P_0 = \$37.75$, $\sigma = 0.28$, APR = 0.07, $S = 40$, and $T = ½$:

$$d_1 = \frac{\ln(37.75/40)}{0.28\times(1/2)^{1/2}} + \frac{(1/2)\times(0.07)}{0.28\times(1/2)^{1/2}} + \frac{0.28\times(1/2)^{1/2}}{2} = -0.016637$$

$$d_2 = -0.016637 - 0.28\times(1/2)^{1/2} = -0.214627$$

From Appendix B, $N(d_1) = N(-0.02) = 0.4920$ and $N(d_2) = N(-0.21) = 0.4168$. Therefore,

$$\text{CALL} = P_0\times N(d_1) - S\times N(d_2)\times e^{-T\times \text{APR}} = 37.75\times(0.4920) - 40\times(0.4168)\times e^{-(½)\times(0.07)} = \$2.47$$

$$\text{PUT} = \text{CALL} + S\times e^{-T\times \text{APR}} - P_0 = 2.47 + 40\times e^{-(½)\times(0.07)} - 37.75 = \$3.34$$

B6. a. Using the Black-Scholes OPM with $P_0 = \$28$, $\sigma = 0.25$, APR = 0.06, $S = 30$, and $T = ½$:

$$d_1 = \frac{\ln(28/30)}{0.25\times(1/2)^{1/2}} + \frac{(1/2)\times(0.06)}{0.25\times(1/2)^{1/2}} + \frac{0.25\times(1/2)^{1/2}}{2} = -0.132189$$

$$d_2 = -0.132189 - 0.25\times(1/2)^{1/2} = -0.308966$$

From Appendix B, $N(d_1) = N(-0.13) = 0.4483$ and $N(d_2) = N(-0.31) = 0.3783$. Therefore,

$$\text{CALL} = P_0\times N(d_1) - S\times N(d_2)\times e^{-T\times \text{APR}} = 28\times(0.4483) - 30\times(0.3783)\times e^{-(½)\times(0.06)} = \$1.54$$

Because $S > P_0$, the Exercise Value = 0, so the Time Premium = \$1.54.

b. Repeating part a with P_0 = \$32,

$$d_1 = \frac{\ln(32/30)}{0.25\times(1/2)^{1/2}} + \frac{(1/2)\times(0.06)}{0.25\times(1/2)^{1/2}} + \frac{0.25\times(1/2)^{1/2}}{2} = 0.623179$$

$$d_2 = 0.623179 - 0.25\times(1/2)^{1/2} = 0.446402$$

From Appendix B, $N(d_1) = N(0.62) = 0.7324$ and $N(d_2) = N(0.45) = 0.6736$. Therefore,

$$\text{CALL} = P_0\times N(d_1) - S\times N(d_2)\times e^{-T\times APR} = 32\times(0.7324) - 30\times(0.6736)\times e^{-(1/2)\times(0.06)} = \$3.83$$

Exercise Value = 32-30 = 2, so the Time Premium = \$1.83.

c. Repeating part a with P_0 = \$34,

$$d_1 = \frac{\ln(34/30)}{0.25\times(1/2)^{1/2}} + \frac{(1/2)\times(0.06)}{0.25\times(1/2)^{1/2}} + \frac{0.25\times(1/2)^{1/2}}{2} = 0.966124$$

$$d_2 = 0.966124 - 0.25\times(1/2)^{1/2} = 0.789347$$

From Appendix B, $N(d_1) = N(0.97) = 0.8340$ and $N(d_2) = N(0.79) = 0.7852$. Therefore,

$$\text{CALL} = P_0\times N(d_1) - S\times N(d_2)\times e^{-T\times APR} = 34\times(0.8340) - 30\times(0.7852)\times e^{-(1/2)\times(0.06)} = \$5.50$$

Exercise Value = 34-30 = 4, so the Time Premium = \$1.50.

B7. a. Using the Black-Scholes OPM with P_0 = \$25, $\sigma = 0.35$, APR = 0.06, S = 25, and T = 9/12:

$$d_1 = \frac{\ln(25/25)}{0.35\times(9/12)^{1/2}} + \frac{(9/12)\times(0.06)}{0.35\times(9/12)^{1/2}} + \frac{0.35\times(9/12)^{1/2}}{2} = 0.300015$$

$$d_2 = 0.300015 - 0.35\times(9/12)^{1/2} = -0.003093$$

From Appendix B, $N(d_1) = N(0.30) = 0.6179$ and $N(d_2) = N(0) = 0.5000$. Therefore,

$$\text{CALL} = P_0\times N(d_1) - S\times N(d_2)\times e^{-T\times APR} = 25\times(0.6179) - 25\times(0.5000)\times e^{-(9/12)\times(0.06)} = \$3.50$$

b. Repeating part a with $\sigma = 0.30$,

$$d_1 = \frac{\ln(25/25)}{0.30\times(9/12)^{1/2}} + \frac{(9/12)\times(0.06)}{0.30\times(9/12)^{1/2}} + \frac{0.30\times(9/12)^{1/2}}{2} = 0.303109$$

$$d_2 = 0.303109 - 0.30\times(9/12)^{1/2} = 0.043301$$

From Appendix B, $N(d_1) = N(0.30) = 0.6179$ and $N(d_2) = N(0.04) = 0.5160$. Therefore,

$$\text{CALL} = P_0\times N(d_1) - S\times N(d_2)\times e^{-T\times APR} = 25\times(0.6179) - 25\times(0.5160)\times e^{-(9/12)\times(0.06)} = \$3.12$$

c. Repeating part a with $\sigma = 0.25$,

$$d_1 = \frac{\ln(25/25)}{0.25\times(9/12)^{1/2}} + \frac{(9/12)\times(0.06)}{0.25\times(9/12)^{1/2}} + \frac{0.25\times(9/12)^{1/2}}{2} = 0.316099$$

$$d_2 = 0.316099 - 0.25\times(9/12)^{1/2} = 0.099593$$

From Appendix B, $N(d_1) = N(0.32) = 0.6255$ and $N(d_2) = N(0.10) = 0.5398$. Therefore,

$$\text{CALL} = P_0\times N(d_1) - S\times N(d_2)\times e^{-T\times\text{APR}} = 25\times(0.6255) - 25\times(0.5398)\times e^{-(9/12)\times(0.06)} = \$2.74$$

B8. Using the Black-Scholes OPM with $P_0 = \$90$, $\sigma = 0.30$, APR $= 0.10$, $S = \$110$, and $T = 3/12$:

$$d_1 = \frac{\ln(90/110)}{0.30\times(3/12)^{1/2}} + \frac{(3/12)\times(0.10)}{0.30\times(3/12)^{1/2}} + \frac{0.30\times(3/12)^{1/2}}{2} = -1.096138$$

$$d_2 = -1.096138 - 0.30\times(3/12)^{1/2} = -1.246138$$

From Appendix B, $N(d_1) = N(-1.10) = 0.1357$ and $N(d_2) = N(-1.25) = 0.1056$. Therefore,

$$\text{CALL} = P_0\times N(d_1) - S\times N(d_2)\times e^{-T\times\text{APR}} = 90\times(0.1357) - 110\times(0.1056)\times e^{-(3/12)\times(0.10)} = \$0.88$$

B9. a. Using the Black-Scholes OPM with $P_0 = \$21$, $\sigma = 0.30$, APR $= 0.04$, $S = 20$, and $T = ½$:

$$d_1 = \frac{\ln(21/20)}{0.30\times(1/2)^{1/2}} + \frac{(1/2)\times(0.04)}{0.30\times(1/2)^{1/2}} + \frac{0.30\times(1/2)^{1/2}}{2} = 0.430346$$

$$d_2 = 0.430346 - 0.30\times(1/2)^{1/2} = 0.218214$$

From Appendix B, $N(d_1) = N(0.43) = 0.6664$ and $N(d_2) = N(0.22) = 0.5871$. Therefore,

$$\text{CALL} = P_0\times N(d_1) - S\times N(d_2)\times e^{-T\times\text{APR}} = 21\times(0.6664) - 20\times(0.5871)\times e^{-(½)\times(0.04)} = \$2.48$$

b. Repeating part a with APR $= 0.07$,

$$d_1 = \frac{\ln(21/20)}{0.30\times(1/2)^{1/2}} + \frac{(1/2)\times(0.07)}{0.30\times(1/2)^{1/2}} + \frac{0.30\times(1/2)^{1/2}}{2} = 0.501057$$

$$d_2 = 0.501057 - 0.30\times(1/2)^{1/2} = 0.288925$$

From Appendix B, $N(d_1) = N(0.50) = 0.6915$ and $N(d_2) = N(0.29) = 0.6141$. Therefore,

$$\text{CALL} = P_0\times N(d_1) - S\times N(d_2)\times e^{-T\times\text{APR}} = 21\times(0.6915) - 20\times(0.6141)\times e^{-(½)\times(0.07)} = \$2.66$$

c. Repeating part a with APR = 0.10,

$$d_1 = \frac{\ln(21/20)}{0.30\times(1/2)^{1/2}} + \frac{(1/2)\times(0.10)}{0.30\times(1/2)^{1/2}} + \frac{0.30\times(1/2)^{1/2}}{2} =$$

$$d_2 = 0.571767 - 0.30\times(1/2)^{1/2} = 0.359635$$

From Appendix B, $N(d_1) = N(0.57) = 0.7157$ and $N(d_2) = N(0.36) = 0.6406$. Therefore,

$$\text{CALL} = P_0\times N(d_1) - S\times N(d_2)\times e^{-T\times\text{APR}} = 21\times(0.7157) - 20\times(0.6406)\times e^{-(1/2)\times(0.10)} = \$2.84$$

B10. Using the Black-Scholes OPM with $P_0 = \$24.38$, $\sigma = 0.48$, APR = 0.06, $S = \$20$, and $T = 9/12$:

$$d_1 = \frac{\ln(24.38/20)}{0.48\times(9/12)^{1/2}} + \frac{(9/12)\times(0.06)}{0.48\times(9/12)^{1/2}} + \frac{0.48\times(9/12)^{1/2}}{2} = 0.792487$$

$$d_2 = 0.792487 - 0.48\times(9/12)^{1/2} = 0.376795$$

From Appendix B, $N(d_1) = N(0.79) = 0.7852$ and $N(d_2) = N(0.38) = 0.6480$. Therefore,

$$\text{CALL} = P_0\times N(d_1) - S\times N(d_2)\times e^{-T\times\text{APR}} = 24.38\times(0.7852) - 20\times(0.6480)\times e^{-(9/12)\times(0.06)} = \$6.75$$

Currently, the Exercise Value = 24.38 - 20 = \$4.38, and the Time Premium = \$2.37.

B11. a. After two periods, the stock price will be either \$250 (with p = 0.55×(0.55) = 0.3025), \$200 (with p = 2×(0.55)×(0.45) = 0.495) or \$150 (with p = 0.45×(0.45) = 0.2025). The exercise value of the call option is \$40 if the stock price is \$250 and zero otherwise. We can compute the value of the call directly using Equation (26.2):

$$\text{CALL} = \left[\frac{1}{(1.10)^2}\right]\times(0.3025\times[40] + 0.495\times[0] + 0.2025\times[0]) = \$10$$

b. The put value can also be computed directly by noting that the exercise value of the put is \$10 if the stock price is \$200, \$60 if the stock price is \$150, and zero if the stock price is \$250:

$$\text{PUT} = \left[\frac{1}{(1.10)^2}\right]\times(0.3025\times[0] + 0.495\times[10] + 0.2025\times[60]) = \$14.13$$

c. To show that Put-Call Parity gives the same answer we got by computing the put value directly in part b, we must first compute the APR with continuous compounding:

$$\text{APY} = 0.10 = e^{\text{APR}} - 1;\ \text{rearranging, } 1.10 = e^{\text{APR}};\ \text{so that, APR} = \ln(1.10) = 0.09531$$

Then Put-Call Parity gives:

$$\text{PUT} = \text{CALL} + S\times e^{-T\times\text{APR}} - P_0 = 10 + 210\times e^{-2\times(0.09531)} - 169.42 = \$14.13$$

B12. a. Yes. Because the interest rate differential on the fixed rate loan (3%) is greater than the interest rate differential on the floating rate loan (1%), it is possible to construct a mutually advantageous interest rate swap.

b. The manufacturing company should turn down the bank's offer. If it borrows at LIBOR + 1%, receives LIBOR from the bank, and pays the bank 11.5%, the net effect is that it pays 12.5%. Since it can borrow on its own at a fixed rate of 12%, it should not accept the swap offer.

c. This swap is advantageous for the bank because it can borrow at 9%, receive 10.5%, and pay LIBOR. The net effect is that it can borrow at LIBOR - 1.5% and save 1.5% relative to the rate it would pay if it borrowed on its own at a floating rate (LIBOR).

d. The bank realizes 3/4 of the net cost benefit and the manufacturing company realizes 1/4.

B13. a. Yes. Because the interest rate differential on the fixed rate loan (3.5%) is greater than the interest rate differential on the floating rate loan (1.9%), it is possible to construct a mutually advantageous interest rate swap.

b. Cajun should not accept the swap. If it does so pays LIBOR + 2% to outside lenders, receives LIBOR from Exxon, and pays 11% fixed to Exxon. The total "fixed" cost to Cajun is 13% versus the 12% at which it can borrow elsewhere.

c. Yes, Cajun should accept this swap. It costs the firm 2% + 9.5% or 11.5% fixed--a savings of 0.5%.

B14. a. Petrie could hedge by buying puts on Toys 'R' Us common stock.

b. Using the Black-Scholes OPM with $P_0 = \$40$, $\sigma = 0.28$, APR = 0.06, $S = 40$, and $T = 1/4$:

$$d_1 = \frac{\ln(40/40)}{0.28\times(1/4)^{1/2}} + \frac{(1/4)\times(0.06)}{0.28\times(1/4)^{1/2}} + \frac{0.28\times(1/4)^{1/2}}{2} = 0.17$$

$$d_2 = 0.17 - 0.28\times(1/4)^{1/2} = 0.03$$

From Appendix B, $N(d_1) = N(0.17) = 0.5675$ and $N(d_2) = N(0.03) = 0.5120$. Therefore,

$$\text{CALL} = P_0\times N(d_1) - S\times N(d_2)\times e^{-T\times \text{APR}} = 40\times(0.5675) - 40\times(0.5120)\times e^{-(1/4)\times(0.06)} = \$2.5249$$

$$\text{PUT} = \text{CALL} + S\times e^{-T\times \text{APR}} - P_0 = 2.5249 + 40\times e^{-(1/4)\times(0.06)} - 40 = \$1.93$$

To buy a \$40 put on 200,000 shares would cost \$386,000 (= 1.93×200,000).

c. Using the Black-Scholes OPM with $P_0 = \$40$, $\sigma = 0.28$, APR = 0.06, $S = 35$, and $T = 1/4$:

$$d_1 = \frac{\ln(40/35)}{0.28\times(1/4)^{1/2}} + \frac{(1/4)\times(0.06)}{0.28\times(1/4)^{1/2}} + \frac{0.28\times(1/4)^{1/2}}{2} = 1.1238$$

$$d_2 = 1.1238 - 0.28\times(1/4)^{1/2} = 0.9838$$

From Appendix B, $N(d_1) = N(1.1238) = 0.8694$ and $N(d_2) = N(0.9838) = 0.8374$. Therefore,

$$\text{CALL} = P_0\times N(d_1) - S\times N(d_2)\times e^{-T\times APR} = 40\times(0.8694) - 40\times(0.8374)\times e^{-(1/4)\times(0.06)} = \$5.9037$$

$$\text{PUT} = \text{CALL} + S\times e^{-T\times APR} - P_0 = 5.9037 + 35\times e^{-(1/4)\times(0.06)} - 40 = \$0.38$$

To buy a \$35 put on 200,000 shares would cost Petrie \$76,000 (= .38×200,000).

d. The at-the-money \$40 put will have value at expiration if Toys 'R' Us declines in price below \$40. The out-of-the-money \$35 will have value at expiration if Toys 'R' Us falls in price below \$34.62 (= 35 - 0.38). For the extra protection of \$5 per share (= 40-35) offered by the \$40 put Petrie must pay an additional \$1.55 (= 1.93 - 0.38) per share. If Toys 'R' Us stock remains essentially unchanged at \$40 or goes up in price, Petrie would have been better-off buying the \$35 put. If the stock declines below \$38.45 (= 40 - 1.55) Petrie would have been better off buying the \$40 put.

B15. a. Three D could buy an interest rate cap or short some futures contracts on short-term, fixed-income securities.

b.

Quarter	1	2	3	4	5	6	7	8
APR	7.50	6.75	8.50	9.00	9.00	9.00	9.00	7.50
Three D's cost ($ millions)	18.75	16.88	21.25	22.50	22.50	22.50	22.50	18.75

c. The seller of the cap must pay Three D whenever the rate rises above 9%.

Quarter	1	2	3	4	5	6	7	8
Seller's Cost ($ millions)	0.00	0.00	0.00	3.13	6.25	8.13	1.88	0.00

B16. Note that the seller of a collar buys a floor and writes a cap. So the seller receives payment when the interest rate falls below the floor and makes payment when interest rates rise above the cap.

Quarter	1	2	3	4	5	6	7	8
8% Floor ($ millions)	1.25	3.13	0.00	0.00	0.00	0.00	0.00	1.25
10% Cap ($ millions)	0.00	0.00	0.00	(0.63)	(3.75)	(5.63)	0.00	0.00

B17. a. To lock-in an interest rate on bonds it intends to sell in a month, Kmart needs to sell long-term, fixed income futures contracts today.

b. To calculate the hedge ratio, we need to calculate the expected price changes for both the Treasury bond futures and Kmart's bonds. If Kmart's new issue rate goes from 9% to 10%, its bonds will fall in value 9.4646, from 100 to 90.5354 [put in n = 60, r = 5.0, PMT = 4.5, and FV = 100; then compute PV = 90.5354]. The T-bond is currently worth 105.1377 [put in n = 40, r = 3.75, PMT = 4.0, and FV = 100; then compute PV = 105.1377], but with an increase in yield to 8.5% would fall in value to 95.2307 [put in n = 40, r = 4.5, PMT = 4.0, and FV = 100; then compute PV = 95.2307], for a decrease of 9.9071. Using Equation (26.9), the hedge ratio is 0.9553 (= 9.4646/9.9071).

c. Using Equation (26.10), the number of contracts is 955 (= 0.9553×[100,000,000/100,000]).

d. A 1% increase in rates costs $9,464,600 (= [100,000,000]0.094646), but is offset by a profit of $9,461,281 (= 955[100,000]0.099071). All but 0.04% is hedged.

e. An increase of 0.5% would put the Treasury bond at par for a decrease in value of 5.1377 (= 105.1377 - 100.00). The hedge ratio would then be 1.8422 (= 9.4646/5.1377), and the number of contracts would be 1842 (= 1.8422[100,000,000/100,000]). A 1% increase in rates costs $9,464,600 (= [100,000,000]0.094646), but is offset by a profit of $9,463,643 (= 1842[100,000]0.051377). All but 0.01% is hedged.

C1. We need to find the value for σ that solves the following equation:

$$\text{CALL} = P_0 \times N(d_1) - S \times N(d_2) \times e^{-T \times \text{APR}} = 32.75 \times N(d_1) - 35 \times N(d_2)\, e^{-(5/12) \times (0.06)}$$

$$2.03 = 32.75 \times N(\text{d}_1) - 34.13585 \times N(\text{d}_2)$$

This equation is solved by trial and error by choosing possible values for σ, computing d_1 and d_2, using Appendix B to find $N(d_1)$ and $N(d_2)$, and calculating the right-hand side of the above equation. This process is repeated to find a call value as close as possible to $2.03 as follows:

σ	d_1	d_2	$N(d_1)$	$N(d_2)$	CALL
0.25	-0.18	-0.34	0.4286	0.3669	1.512
0.30	-0.12	-0.31	0.4522	0.3783	1.896
0.35	-0.07	-0.30	0.4721	0.3821	2.420
0.32	-0.10	-0.30	0.4602	0.3821	2.028

σ = 32% yields a value for the call of $2.028, which we deem "close enough."

C2. Consider the cost of buying an asset, buying a put with a strike price of *S*, and selling a call with a strike price of *S*. The asset will cost P_0, the put will cost PUT and you will receive CALL for the call. As a result, your total expenditure is P_0 + PUT - CALL. Solving the Put-Call Parity equation for this expression gives: $P_0 + \text{PUT} - \text{CALL} = S \times e^{-T \times \text{APR}}$. But $S \times e^{-T \times \text{APR}}$ is simply the present value of *S* dollars to be received in *T* periods when the discount rate is the continuously-compounded riskless rate of return, APR. As a result, buying an asset, buying a put, and selling a call is equivalent to buying a *T*-period riskless bond that pays *S* dollars when it matures.

C3. With the convertible bond, the call option cannot be exercised without relinquishing the bond. The market value of the warrant-bond combination would be expected to be somewhat higher than the market value of a comparable convertible bond because the warrant-bond combination provides the additional option of exercising the warrant while keeping the bond, and we know from the Options Principle that this additional option has a non-negative value.

C4. When there is no chance that the option will be out-of-the-money in the remaining time until maturity, the value of the European put is the present value of the exercise value, $(S - P_0) \times e^{-T \times APR}$. In contrast, because the American option can be exercised now, its value is $S - P_0 > (S - P_0 \times) e^{-T \times APR}$ for any $T > 0$.

C5. For an out-of-the-money option, the only value for the option is due to the possibility that it will be in-the-money before it expires, and the closer the value of the underlying asset is to the strike price, the greater the probability that the option will become in-the-money before its expiration. Thus, the time premium of an out-of-the-money option increases (decreases) as the value of the underlying asset moves toward (away from) the strike price. For an in-the-money option, the time premium is due to a combination of (1) the possibility that it will be even deeper in-the-money before it expires, and (2) the forgone opportunity of earning the time value of money on the exercise value (that is, exercising the option now and reinvesting the proceeds). As the value of the underlying asset moves farther from the strike price, the forgone time value of money on the exercise value grows, but the value of the incremental gain from becoming deeper in-the-money does not grow. Thus, the time premium of an in-the-money option decreases (increases) as the value of the underlying asset moves away from (toward) the strike price.

C6. Consider a swap from the perspective of a party who is paying fixed in exchange for floating. After each settlement date the party is obligated to sell a fixed rate cash flow at the next settlement date. This obligation is like a forward contract on interest rates. Since a swap will involve a number of such settlements it is like a portfolio of forwards.

C7. Futures contracts call for daily settlement. At the close of each trading day, funds are transferred from the accounts of that day's losers in the futures market to the accounts of that day's winners. Implicit forward contracts are then written for the next day.

CHAPTER 27 -- BANKRUPTCY, REORGANIZATION, AND LIQUIDATION

A1. A firm is in **financial distress** when it is unable to pay its debts as they come due.

A2. Indicators of impending financial distress include: accounting losses; interest coverage worsening; the debt-to-equity ratio increasing; operations consuming more cash than they generate; and net working capital turning negative.

A3. The primary purpose of the bankruptcy code is to enable creditors to recover as much of their loans as possible. A financially distressed but viable firm should file for Chapter 11 when it is or will soon be unable to pay its debts when they come due and it cannot reach agreement with its creditors about restructuring its debt outside of court.

A4. Under a Chapter 11 reorganization, a plan is developed to reorganize the debtor's business and restore its financial health. Under a Chapter 7 liquidation, the assets of the firm are sold.

A5. Managerial incompetence is by far the leading cause of financial distress. One survey indicated that incompetence caused 94% of business failures. However, external factors can also cause financial distress. Intense competition, credit squeezes, and oil embargoes are a few examples of external factors which might send some firms into distress.

A6. Young firms are more likely to fail than older firms. See Table 27-3 for a breakdown by age of business failures.

A7. The basic premise of Chapter 11 reorganization is that the debtor should be allowed to attempt reorganization provided the going-concern value of the debtor exceeds its liquidation value.

A8. a. A **debtor-in-possession** is a firm which has filed a petition for relief under Chapter 11.

b. An **automatic stay** is a fundamental protection provided to debtors under the bankruptcy code. It halts further efforts by creditors to collect their debts or seize collateral.

c. A **plan of reorganization** is the plan the firm files under Chapter 11 which specifies a new business plan and capital structure as well as how old debts will be paid and cash or new securities distributed.

d. In a **consensual plan of reorganization**, all classes of creditors and shareholders must vote to accept the reorganization plan.

e. In a **cramdown**, the court mandates the acceptance of a plan in spite of dissent.

f. A **prepackaged bankruptcy** involves the debtor and creditors negotiating a plan of reorganization and then filing it along with the bankruptcy petition.

A9. The cramdown provision reduces the incentive for some classes of creditors to holdout.

A10. Prepackaged bankruptcy is most effective when there is only one class of creditors that will be disadvantaged by the plan, and when the debtor can obtain approval of that class before filing.

A11. a. Reorganization is preferred to liquidation when the firm's reorganization value exceeds the firm's liquidation value.

b. Liquidation is preferred to reorganization when the firm's liquidation value exceeds its reorganization value.

A12. No, absolute priority is not always observed. It may be in the best interests of the senior creditors to make a small distribution to junior classes in order to accelerate the procedure and reduce expenses.

A13. It is highly unlikely that any one variable will predict bankruptcy across all industries and times.

A14. Positive incentives to participate include offering to exchange new securities with higher interest rates, shorter maturities, senior rankings, or stronger covenants, than the outstanding securities. Negative incentives include the theat of filing for bankruptcy and of structuring the exchange offer so that nonexchanging holders end up holding securities junior to the new securities.

A15. **Holdouts**are holders of outstanding securities who refuse to exchange their securities for the new ones being offered. They frustrate the reorganization process by delaying or even preventing the reorganization from going through.

A16. An out-of-court restructuring is usually less disruptive to the debtor's business and less time-consuming than the bankruptcy process. On the other hand, bankruptcy has advantages relative to out-of court restructuring. Filing the petition stays all creditor collection efforts, and bankruptcy provides a single forum for resolving disputes. In addition, the bankruptcy court can authorize debtor-in-possession financing to relieve a liquidity crisis and can authorize the debtor to reject unfavorable leases and other contracts.

A17. The 5 basic conditions that a plan of reorganization must satisfy are:
(1) the plan must be feasible;
(2) the plan cannot discriminate unfairly among creditors of equal classes;
(3) the plan must be accepted by at least one class of creditor;
(4) the plan must satisfy the fair and equitable test; and
(5) the plan must satisfy the best interests of creditors test.

A18. a. To satisfy the **best-interests-of-creditors test** a plan must provide dissenting creditors at least as much as if the debtor were liquidated.

b. To satisfy the **fair-and-equitable test** (1) a plan must have been accepted by a class of claims, or else (2) the class's claims must be paid in full, or else (3) if the class rejects the plan and is not paid in full, then no class junior to that class can receive anything.

B1. The **absolute priority** doctrine holds that creditors should be compensated for their claims in accord with a rigid hierarchy and that more senior claims should be paid in full before more junior claims receive anything. In principle, adhering to absolute priority both allows creditors to know where they stand and provides creditors what they should receive. However, senior claimants often find that it expedites the process and reduces expenses to compensate junior claimants.

B2. Chapter 10 was intended to allow small and medium-sized businesses to reorganize under a "fast track Chapter 11" procedure, which was designed to reduce the delay and cost of Chapter 11.

B3. In order to determine whether a firm should be reorganized or liquidated, both reorganization and liquidation values need to be estimated. In addition, estimates of the values of their claims are of critical importance to all claimants in order for them to know how to vote on proposed plans and what to expect if a plan is accepted.

B4. The main advantages of a prepackaged bankruptcy are: (1) it can alleviate the holdout problem; (2) once confirmed, it is binding on all debtholders; (3) it permits the debtor to realize tax advantages not available in an out-of-court restructuring; and (4) it allows the debtor to reject burdensome leases and other contracts. Not all firms can take advantage of prepackaged bankruptcy. Firms with multiple classes of creditors may have difficulty getting all the acceptances they need.

B5. a. Using Equation (27.1), the Z-score is:

$$Z = 0.012\times(5) + 0.014\times(10) + 0.033\times(-5) + 0.006\times(50) + 0.999\times(1.05) = 1.43$$

b. Since $Z < 1.81$, this firm is likely to go bankrupt within the year.

B6. a. Using Equation (27.1), the Z-score is:

$$Z = 0.012\times(5) + 0.014\times(20) + 0.033\times(2) + 0.006\times(110) + 0.999\times(1.25) = 2.36$$

b. Since $1.81 \le Z \le 2.99$, it is difficult to determine whether this firm is likely to go bankrupt within the year.

B7. Firm and industry characteristics change over time, which results in the discriminant coefficients changing.

B8. This is a bond valuation problem. Using your calculator, set $n = 10$, $r = 16$, PMT = 100, FV = 1000, and solve for PV = 710. So, the new debt is worth 71% of its face value.

B9. In a prepackaged bankruptcy, the debtor and creditors negotiate a plan of reorganization and only then file a bankruptcy petition. In a private restructuring, the debtor and creditors negotiate a restructuring/reorganization, but don't file a petition.

B10. In order of decreasing priority, the claims are:

1. Legal fees incurred during the bankruptcy process
2. First mortgage bonds
3. Bank debt secured by inventories and receivables
4. Second mortgage bonds
5. Unsecured bank debt
6. Accounts payable
7. Subordinated debentures
8. Preferred stock
9. Common stock

B11 a. There are $800 million in secured claims but only $600 million in assets. The banks should receive both their 5/8 share plus that of the subordinated debentures' share of 1/8, for a total of $450 million (= 600×[6/8]). The trade creditors should receive the balance of $150 million (= 600×(2/8]). The plan discriminates against the banks.

b. The plan does not pass the fair-and-equitable test, because the banks are not being paid in full but a junior claim, the debenture, is receiving some payment.

B12. a. Based on the information in B11-a. above, the trade creditors are being unfairly treated. They should receive $150 million not $100 million.

b. Probably not, the plan does not provide the trade creditors with an amount equal to their unsecured claims.

B13 No one is discriminated against. The banks are due only the amount needed to pay them in full. Given the subordination of the debentures, it might appear that the banks should get $525 million (= 700×[6/8]), but this is in excess of the $500 million owed them. The debenture holders get this excess.

B14. If Olympic Computer's subordinated debt is subordinated to the accounts payable and the mortgage bonds, as well as to the bank debt, the subordinated creditors will not receive anything unless the holders of the accounts payable *and* the bank debt *and* the mortgage bonds are *all* paid in full. These unsecured claims sum to $6 million (= 4 + 1 + 1) but there are only $3 million of assets available to satisfy them. Each unsecured claim other than the subordinated debt thus receives 50 cents (= 3/6) on the dollar so that the distributions are:

	Secured claim	Unsecured claim	Total distribution
Trade creditors	--	$2.0 (= 4×[3/6])	$2.0
Banks	$2.0	0.5 (= 1×[3/6])	2.5
Subordinated debtholders	--	--	--
Mortgage bondholders	1.0	0.5 (= 1×[3/6])	1.5
Preferred stockholders	--	--	--
Common stockholders	--	--	--
	$3.0	$3.0	$6.0

B15. The secured mortgage holders receive the $50 million from the sale of the new factory. That leaves them with $50 million unsecured. There are a total of $250 million (= 50 + 200) in unsecured claims outstanding which are backed by $150 million in assets. The mortgage holders receive an additional $30 million (= 150×[50/250]), for a total of $80 million. The other unsecured creditors receive $120 million (= 150×[200/250]).

C1. No, not necessarily. The Act was designed to simply filing for reorganization-type procedures, so if the Act achieved its purpose, we should expect to see more bankruptcy filings. In addition, based on recent evidence, the United States ranks well behind many Western European countries in bankruptcies per capita.

C2. The types of financial intermediaries that have been growing the most rapidly in recent years (for example, pension funds and mutual funds) typically have multiple clientele to satisfy. These intermediaries hold multiple types of securities on behalf of their various cliental in the same firm. To attempt to satisfy (or placate) their different cliental, intermediaries will probably want relative rather than absolute priority rules enforced.

C3. a. Yes, a firm which has been experiencing large losses over a 5-year period is likely to have productive assets which are obsolete or no longer allow it to remain competitive. In addition, a firm which has been losing money regularly may well have over-valued inventories and receivables which are in jeopardy. These problems are intensified when the firm is in a high technology industry like personal computer manufacturing.

b. If we assume that the mortgage bonds are backed by the plant and equipment, then the mortgage bond holders will be paid in full. That leaves $49 million (= 10 + 16 +16 + 5 + 2) to pay the $100 million in senior unsecured claims. The latter claims will receive $.49 on the dollar or $24.5 million for the banks, $14.7 million for the trade creditors, and $9.8 million for the senior debenture holders. The subordinated debenture holders and stockholders would receive nothing.

c. In the plan outlined above, the mortgage bondholders would go home happy. However, the senior unsecured claimholders would probably argue that their funds were implicitly backed by the plant and equipment and that they should receive more than $5 million of the plant and equipment's value.

C4. a. The bankruptcy code imposes an *automatic stay*, which halts any further efforts by the debenture holders and the secured creditor to collect their debts or by the secured creditor to arrange for a sheriff's sale of the property on which it has the lien.

b. The secured debt will have its maturity reinstated; the secured creditor will receive its interest when due; it will receive overdue interest along with a penalty; and its lien will continue. The class of secured debt thus appears unimpaired; the secured creditor's position is restored to what it was before the bankruptcy filing. Provided the other requirements in Table 27-5 for confirming the plan are met, the bankruptcy court can confirm the plan despite the secured creditor's objection to it.

c. Assuming that the other 99% of the holders, who own 75% of the debentures, have voted in favor of the plan, the class of debentures has approved the plan along with the other two classes of securityholders. Provided the bankruptcy court finds that the objecting debenture holder would get at least as much as in a liquidation (the best-interests-of-creditors test), the bankruptcy court can confirm the plan.

The best-interests test is likely met because in a liquidation the secured creditor would receive the value of its lien, $3 million (- 0.5×6 million), plus its pro rata share of what's left over. There are $4.5 million (= 5 - 3 + 2.5) of unsecured claims but only $3 million (= 6 - 3) worth of assets. The secured creditor would get $1.33 million (= [2/4.5]×3), leaving $1.67 million (= 3 - 1.33) for the debenture holders. Even at a 20% discount rate, the present value of the six annual installments of $600,000 each equals $1.995 million, which exceeds $1.67 million.

d. If all the debenture holders object, the class rejects the plan. The court nevertheless can confirm the plan under the cramdown procedure but only if it finds that the present value of the six annual installments of $600,000 each exceeds the sum of $2.5 million plus accrued and unpaid interest, which is the amount of the debenture holders' claim. For example, assuming a 10% APR and one year's unpaid interest, the amount of the claim is $2.75 million (= 2.5 + 0.10×2.5). In that case, the required return would have to be no greater than 8.28% [put in n = 6, PV = 2.75, PMT = 0.6, and FV = 0; then compute r = 8.28%]. For example, if the required return is 15%, the present value is $2.27 million [put in n = 6, r = 15, PMT = 0.6, and FV = 0; then compute PV = 2.27], which is less than the debenture holders' claim. So the plan can't be crammed down over their objection, because they would receive less than 100% of their claim while the common stockholders, who are junior to them, receive the value of the common stock.

e. DISSIPATED TECHNOLOGIES CORPORATION

Balance Sheet before Reorganization ($ millions)

Assets		Liabilities and Stockholders' Equity	
Total assets	$6.0	Secured debt	$5.0[a]
		Debentures	2.5[a]
		Common stock	(1.5)
		Total liabilities plus stockholders' equity	$6.0

[a]Excluding accrued interest.

DISSIPATED TECHNOLOGIES CORPORATION

Balance Sheet after Reorganization ($ millions)

Assets		Liabilities and Stockholders' Equity	
Total assets	$6.0	Secured debt	$3.63[c]
		Debentures	2.27[a]
		Common stock	0.10[b]
		Total liabilities plus stockholders' equity	$6.00

[a]Payments discounted at 15% APR.
[b]Assigned a nominal value of 0.10 because the common stockholders have the most junior claim.
[c]Calculated as the difference between the value of the assets (6.0) and the value of the junior claims (2.37).

CHAPTER 28 -- MERGERS AND ACQUISITIONS

A1. a. A **merger** involves the combination of two firms in which one firm (the acquiror) absorbs all the assets and liabilities of the other firm (the acquiree) and assumes the acquiree's businesses.

b. A **consolidation** occurs when two or more firms combine to form an entirely new entity.

c. A **horizontal merger** involves two firms in the same line of business.

d. A **vertical merger** involves integrating forward toward the consumer or backward toward the source of supply in a particular line of business.

e. A **conglomerate merger** involves firms in unrelated businesses.

A2. The valid motives two firms may have for merging are: (1) to achieve operating efficiencies by eliminating duplicate facilities, operations, or departments or to achieve economies of scale in production, distribution, or some other phase of their operations; (2) to realize tax benefits; (3) to access surplus cash; and (4) to grow more quickly or more cheaply than is possible with internal development.

A3. Diversification is usually an invalid justification for conglomerate mergers because individual investors are able to diversify their holdings more cheaply on their own.

A4. Financial synergy results when the merged firm is able to make larger securities issues and, thus, reduce its transaction costs relative to the unmerged firms. Business synergy results from eliminating duplicate facilities, operations, or departments and from achieving economies of scale in the operation of the merged firm. Financial synergy is a function of the way the firm is financed, and business synergy is a function of the way the firm is operated; business synergy is likely to be much more important than financial synergy is creating value through a merger.

A5. a. $V_{AB} = V_A + V_B + \text{PV(savings)} = 50 + 75 + 10 = \$135 \text{ million}$

b. By Equation (28.1),

$$\text{NAM} = V_{AB} - (V_B + P_B) - E - V_A = 135 - (75 + 0) - 5 - 50 = \$5 \text{ million}$$

So there is a net advantage to merging in this case.

c. $$\text{NAM} = V_{AB} - (V_B + P_B) - E - V_A = 135 - (75 + 3) - 5 - 50 = \$2 \text{ million}$$

Because the NAM > 0, the acquisition would be advantageous to Ace's shareholders. Ace's shareholders receive the \$2 million net advantage, and Brace's shareholders receive the \$3 premium.

A6. a. With a P/E ratio of 20 and EPS of \$5, the market value of the acquiror's stock is \$100 (= 20×5) per share. Similarly, the market value of the acquiree's stock is \$20 per share. With no premium, the acquiror must exchange 1/5 of a share for each share of the acquiree's stock.

b. The acquiror will need to issue an additional 1 million shares (= [1/5]× 5 million) in exchange for the acquiree's stock. The combined earnings of the merged firm will be $60 million (= 5/share×[10 million shares] + 2/share×[5 million shares]). The resulting EPS for the acquiror will be $5.45 (= 60 million/11 million shares).

c. The new P/E ratio for the acquiror is 18.35 (= 100/5.45).

d. Merging with a firm that has a lower P/E ratio than the acquiror will lead to an increase in the acquiror's EPS (when no premium is paid); however, the acquiror's P/E ratio falls. The increased earnings per share and decreased P/E ratio reflect the accounting impact from merging with a firm that has a lower P/E ratio rather than any true economic gains.

A7. The net advantage to merging typically goes almost entirely to the shareholders of the acquiree with the shareholders of the acquiror receiving none of it, or only a modest portion, on average.

A8. The three basic ways of effecting a corporate acquisition are: (1) a merger or consolidation, in which the acquiring entity (a new entity in the case of consolidation) obtains all the assets and assumes all the liabilities of the acquired firm(s); (2) a purchase of stock, in which the acquiror purchases the acquiree's stock directly from the acquiree's shareholders; and (3) a purchase of assets, in which the acquiror purchases only the selling firm's assets (and perhaps assumes certain specified liabilities). In a merger or consolidation, the acquired firm loses its corporate existence, and shareholder approval is normally required by the shareholders of both firms involved. The purchase of stock does not require shareholders' meetings; the shareholders of the acquiree effectively express their approval of the transaction by agreeing to sell their shares to the acquiror.

A9. In a **tax-free acquisition**, the selling shareholders are treated as having exchanged their old shares for substantially similar new shares and the acquiring corporation's tax basis in each asset whose ownership is transferred is the same as the acquiree's. In a **taxable acquisition**, the selling shareholders are treated as having sold their shares and the acquiror can, if it chooses, increase the tax basis in the assets it acquires to their respective fair market values. The seller generally prefers a tax-free acquisition because the selling shareholders who receive only stock do not have to pay any tax on the gain until the shares are sold. The purchaser generally prefers a taxable transaction because of the ability to write up depreciable assets for tax purposes.

A10. To qualify as a tax-free acquisition, the transaction must have a sound business purpose and not be solely for tax reasons. The acquiror must continue to operate the acquiree's business and meet certain requirements regarding the mode of acquisition and medium of payment.

A11. Under the **pooling-of-interests** method the respective assets, liabilities, and operating results of the merging firms are added together without adjustments. Under the **purchase** method, the acquiror is treated as having purchased the assets of the other firm, and the purchase price, after adding the fair market value of the liabilities the acquiror assumes, is allocated to the assets acquired. Any excess of the purchase price over the fair market value of the net assets acquired is recorded as goodwill. In an efficient capital market, the choice of accounting technique should not affect market value because it does not affect cash flows.

A12. A **tender offer** involves an offer to purchase shares of stock at a stated price from target firm shareholders who are willing to sell their shares at that price. A firm can acquire another firm through a tender offer by purchasing enough shares to gain effective control of the target firm. The principal advantages of a tender offer relative to other acquisition methods are: (1) it is the quickest means of obtaining control of another firm; (2) it offers greater flexibility than other methods; (3) it is the simplest way for a foreign firm to buy an American firm; and (4) open market purchases of the target firm's stock followed by a cash tender offer give the potential acquiror an opportunity for profit.

A13. A **proxy contest** occurs when one or more individuals opposed to incumbent management solicit target shareholders' proxies to vote at the next annual stockholder meeting in order to (try to) elect a dissident slate of directors. A firm can acquire another firm through a proxy contest by securing enough votes to elect its own board of directors. The principal disadvantages of a proxy contest relative to a cash tender offer are that proxy fights are expensive and time-consuming and few succeed.

A14. The principal defensive tactics that firms have used in an effort to fight off unwanted suitors are of two major types. Anticipatory tactics include: (1) dual class recapitalizations--issuing a second class of common stock with superior voting rights; (2) employee stock ownership plans--the firm sells a large block of stock to the ESOP and repurchases an equivalent number of shares in the open market and the firm votes the ESOP's shares until they are distributed to employees; (3) poison pills--giving its shareholders the right to buy its own shares at half price prior to a merger or to buy acquiror shares at half price following a merger; (4) staggered election of directors--this prevents a hostile raider from gaining control in a single proxy contest; and (5) supermajority voting--requiring a supermajority of the firm's common shareholders to approve a merger that has not been approved by the board of directors. In addition, there are responsive tactics such as: (1) selling assets the acquiror wants or buying assets the acquiror does not want; (2) leveraged recapitalization--this raises the leverage ratio of the target and makes it more difficult for the raider to make the acquisition with borrowed funds; (3) litigation; (4) counter tender offers for the common stock of the raider firm; (5) large scale repurchases to eliminate excess cash or "greenmail" repurchases to eliminate dissident shareholders; and (6) standstill agreements that prevent a potential raider from acquiring more than a stated percentage of the firm and that require the potential raider to vote its shares with management.

Defensive tactics would generally be expected to have a negative impact on shareholder wealth. Defensive tactics provide mechanisms for entrenching existing management and may allow those entrenched managers to take actions in their own interest to the detriment of the firm's shareholders.

A15. A **leveraged buyout** is distinguished from other types of acquisitions because it is financed principally, sometimes more than 90%, by borrowing on a secured basis.

B1. a. Analyzing the potential merger as we would any other capital investment project, we find:

$$NPV = -100 + \frac{10}{(1.20)} + \frac{20}{(1.20)^2} + \frac{30}{(1.20)^3} + \frac{40}{(1.20)^4} + \frac{250}{(1.20)^5} = \$59.3 \text{ million}$$

b. The internal rate of return is the discount rate that makes the NPV zero. IRR = 34.45%.

c. Payback = 4 years.

d. If the cash flow and net sales estimates are reasonable (not overly optimistic), then Firm A should proceed with the merger in order to realize the $59.3 million net present value.

B2. Because the merger effectively results in each firm guaranteeing the debt of the other firm, the probability of bankruptcy is decreased and the expected cash flows to bondholders is increased. Thus, bondholder wealth is increased and stockholder wealth is decreased. This happens in any stock-for-stock transaction where the two firms' operating cash flows are less than perfectly correlated.

B3. a. Firm A is worth:

$0.3\times(200) + 0.5\times(100) + 0.2\times(25) = \$115$ million.

Firm B is worth:

$0.3\times(50+100) + 0.5\times(50+50) + 0.2\times(25+0) = \$100$ million.

Note that the values of firm A and B also appear in the table below.

b.

	Boom	Stable Growth	Recession	Total Market Value
Probability	0.3	0.5	0.2	
Firm A				
Debt	$ -	$ -	$ -	$ -
Equity	200	100	25	115
Total	$ 200	$ 100	$ 25	$ 115
Firm B				
Debt	$ 50	$ 50	$ 25	$ 45
Equity	100	50	-	55
Total	$ 150	$ 100	$ 25	$ 100
Combined				
Debt	$ 50	$ 50	$ 50	$ 50[a]
Equity	300	150	-	165[b]
Total	$ 350	$ 200	$ 50	$ 215

[a]Bondholder wealth increases by $5 (= 50 - 45) because they get paid in full in each state of nature.
[b]Shareholder wealth decreases by $5 (= 115 + 55 - 165) due to the coinsurance effect.

As displayed in the table above, the combined firm has a value of $215 million, with debt valued at $50 million and equity valued at $165 million.

c. No, the firms should not merge. The merger creates a wealth transfer from shareholders to bondholders.

B4. a. As was shown in problem B3, a merger between the two firms insures that bondholders get paid in all states of nature even when there is no synergistic benefit to merging. In this problem, synergy creates a net advantage to merging of $28 million (= 30 - 2), so the bondholders again get paid in all states and the value of debt is $50 million..

b. The total value of the combined firm is $243 million (= 215 + 28)--the sum of the value of the combined firm without synergistic benefits plus the net advantage to merging.

The value of equity of the combined firm is $193 million (= 243 - 50). The value of equity without a merger is, from B3, $170 million. Thus, shareholders benefit from the merger.

c. The firms should merge. The merger increases shareholder wealth. Note that the bondholders are also made better off. Both groups of stakeholders benefit, because they split the synergistic benefit to merging. Shareholders receive $23 million of the benefit and bondholders receive $5 million.

B5. Some shareholders of the acquiree might incur an income tax obligation even though the acquisition qualifies under the Internal Revenue Code as a tax-free acquisition because up to just under 50% of the aggregate purchase price of acquiree can be obtained through a cash tender offer. Those shareholders who receive cash are taxed on the gain immediately while those shareholders who receive only stock defer the tax obligation. In a taxable transaction, shareholders of the acquiree will incur either a tax obligation or a tax credit depending on whether they sell their shares for more or less than they paid for them.

B6. Goodwill = [Purchase Price + MV(liabilities)] - MV(assets) = (275 + 30) - 180 = $125 million.

B7. a. Multiple of earnings paid is 10 (= 50/5).

b. Multiple of cash flow paid is 8 (= 50/6.25).

c. Multiple of book value paid is 2.5 (= 50/20).

d. Premium paid is 43% (= (50-35)/35).

B8. a. The barrels of oil equivalent in International's reserves is 35 million barrels (= 25 million + [60 billion/6000]).

b. Total value of assets is $350 million (= 8.50×[35 million] + 52.5 million).

B9. a. With a 50% premium for Windy City's shares, Razorback will have to pay $45 per share for a total of $225 million (= 45×[5 million]); this is the cost of purchasing the target's common shares. The Windy City debt will require $2.4 million (= 50 M×[0.08]×[1 - 0.4]) in annual after-tax interest payments and, with an after-tax cost of new debt of 6% (= [1 - 0.4]×10%), the PV of the debt displaced is given by the debt service parity approach:

$$PV = \sum_{t=1}^{6} \frac{2.4M}{(1.06)^t} + \frac{50M}{(1.06)^6} = \$47.05 \text{ million}$$

Because this amount is less than the cost of retiring that debt, Razorback should assume the debt, and the net acquisition cost is given by Equation (28.11):

$$\text{NAC} = 225 + 47.05 + 3 = \$275.05 \text{ million}$$

Razorback's target debt-to-equity ration is 1/3 so that its target debt-to-value ratio is 1/4. The added debt that Razorback can issue is given by Equation (28.12):

$$\text{Added Debt} = \text{NAC}(L) - \text{PV(debt assumed)} = 275.05(1/4) - 47.05 = \$21.71 \text{ million}$$

To find the terminal value on a non-disposition basis, we first need to compute the after-tax interest expense. After the acquisition, Razorback will have $50 million of 8% debt and $21.71 million of 10% outstanding; thus, the after-tax annual interest expense is $3.7026 million (= [1 - 0.4]×[0.08×(50 M) + 0.10×(21.71 M)]). Deducting the after-tax interest expense from the year 6 net operating profit and multiplying by the current price-earnings multiple of 10 gives the terminal value of the equity, TV(E) = $712.97 million (= 10×[75 - 3.7026]). The terminal value of the debt is the sum of the debt assumed and the added debt; TV(D) = $71.71 million. Applying Equation (28.13), we can compute the terminal value of the net assets:

$$\text{TV(NA)} = \text{TV(E)} + \text{TV(D)} - T - E = 712.97 + 71.71 - 0 - 0 = \$784.68 \text{ million}$$

Computing the NPV at 16.5% gives:

$$\text{NPV} = -275.05 + \frac{25}{(1.165)} + \frac{30}{(1.165)^2} + \frac{35}{(1.165)^3} + \frac{40}{(1.165)^4} + \frac{45}{(1.165)^5} + \frac{50 + 784.68}{(1.165)^6}$$

$$= \$167.19 \text{ million}$$

The IRR is the discount rate that makes the NPV zero. On a non-disposition basis, the IRR is 27.75%.

b. To compute the terminal value on a disposition basis, the earnings after interest and taxes are multiplied by the acquisition multiple Razorback expects to pay for Macrohard. The current price of $30 per share and price-earnings multiple of 10 implies that the current earnings are $3 per share. The acquisition multiple is, therefore, 15 (= 45/3). Using this multiple gives a terminal equity value of $1,069.46 million (= 15×[75 - 3.7026]). If Razorback sells its Macrohard stock, it will realize a long-term capital gain of $841.46 million (= 1069.46 - 228), and the tax on that gain will be $336.58 million (= 0.4×[841.46]). Again applying , Equation (28.13) we can compute the terminal value of the net assets:

$$\text{TV(NA)} = \text{TV(E)} + \text{TV(D)} - T - E = 1069.46 + 71.71 - 336.58 - 0 = \$804.59 \text{ million}$$

Computing the NPV at 16.5% gives:

$$\text{NPV} = -275.05 + \frac{25}{(1.165)} + \frac{30}{(1.165)^2} + \frac{35}{(1.165)^3} + \frac{40}{(1.165)^4} + \frac{45}{(1.165)^5} + \frac{50 + 804.59}{(1.165)^6}$$

$$= \$175.16 \text{ million}$$

The IRR on a disposition basis is IRR = 28.16%.

c. Yes, Razorback should make the acquisition if it has to pay $45 per share; calculating the terminal value both on a disposition basis and on a non-disposition basis indicates a positive NPV for the merger.

d. The maximum price that Razorback can afford to pay on a disposition basis results in a net acquisition cost such that the NPV of the acquisition is zero:

$$0 = -\text{NAC} + \frac{25}{(1.165)} + \frac{30}{(1.165)^2} + \frac{35}{(1.165)^3} + \frac{40}{(1.165)^4} + \frac{45}{(1.165)^5} + \frac{50 + 804.59}{(1.165)^6}$$

Solving this equation gives NAC = \$450.21 million. Deducting the PV(debt assumed) and the transaction costs gives the maximum cost of purchasing the target's common shares, \$400.16 million (= 450.21 - 47.05 - 3). This corresponds to a maximum price of \$80.03 per share (= 400.16/5).

B10. a. The unleveraged beta is $\beta_U = (1 - L)\times \beta_L = (3/4)\times(1.25) = 0.938$; the unleveraged cost of equity capital for Razorback is $r = r_f + \beta_U\times(r_f - r_m) = 10 + 0.938\times(8) = 17.50\%$.

b. The APV on a non-disposition basis is computed by applying Equation (28.18). To use this equation, we need the net-benefit-to-leverage factor, T^*, and the before-tax interest payments, INT_t. The before-tax interest payments consist of the interest on the assumed debt plus the interest on the added debt:

$$\text{INT}_t = 0.08\times(50 \text{ million}) + 0.10\times(21.71 \text{ million}) = \$6.171 \text{ million}$$

We can solve Equation (16.9) for T^*:

$$\text{WACC} = r - T^*\times L r_d \times\left[\frac{1+r}{1+r_d}\right]$$

$$0.165 = 0.175 - T^*\times\left(\frac{1}{4}\right)\times(0.10)\times\left[\frac{1.175}{1.10}\right]$$

$$T^* = 0.3745$$

Finally, applying Equation (28.15), the net APV is:

$$\text{Net APV} = -\text{NAC} + \sum_{t=1}^{n}\frac{\text{CFAT}_t}{(1+r)^t} + \sum_{t=1}^{n}\frac{T^*\times\text{INT}_t}{(1+r_d)^t}$$

$$= -275.05 + \frac{25}{(1.175)} + \frac{30}{(1.175)^2} + \frac{35}{(1.175)^3} + \frac{40}{(1.175)^4} + \frac{45}{(1.175)^5}$$

$$+ \frac{834.68}{(1.175)^6} + \sum_{t=1}^{6}\frac{0.3745\times(6.171)}{(1.10)^t}$$

$$= \$157.84 \text{ million}$$

c. Again applying Equation (28.15), the APV is

$$\text{Net APV} = -275.05 + \frac{25}{(1.175)} + \frac{30}{(1.175)^2} + \frac{35}{(1.175)^3} + \frac{40}{(1.175)^4}$$

$$+ \frac{45}{(1.175)^5} + \frac{854.59}{(1.175)^6} + \sum_{t=1}^{6} \frac{0.3745 \times (6.171)}{(1.10)^t}$$

$$= \$165.41 \text{ million}$$

B11. a. Using the pooling-of-interests method results in the following balance sheet.

	Combined Firm
Assets (millions)	
Working capital	$ 35
Fixed Assets	150
Total	$185
Liabilities and Equity (millions)	
Debt	$ 55
Equity	130
Total	$185

b. The purchase method of accounting gives the following balance sheet for the combined firm..

	Ocean	Eastern	Adjustments	Combined Firm
Assets (millions)				
Working capital	$ 10	$ 25	+2[a]	$ 37
Fixed Assets	25	125	+7	157
Goodwill			+10[b]	10
Total	$35	$150		$204
Liabilities and Equity (millions)				
Debt	$ 5	$ 50	-1	$ 54
Equity	30	100	-30[c]	
			+50[d]	150
Total	$35	$150		$204

[a]Working capital, fixed assets, and debt are all adjusted to their respective market values.
[b]Goodwill is calculated as the price paid for Ocean's equity minus the market value of its net assets plus the market value of its liabilities, or $10 million (= 50 - (12 + 32) + 4).
[c]The acquisition eliminates Ocean's equity.
[d]The cash paid for Ocean.

B12. a. The price/earnings ratio for Firm A is 10 (=25/2.50).

The price/earnings ratio for B is 12.5 (=50/4).

b. To acquire A, Firm B must issue one share of its own stock for each two shares of A. Thus, Firm B issues 500,000 new shares and has 2,500,000 shares outstanding after the merger.

The combined earnings after the merger are \$10.5 million (=2.5×1,000,000 + 4×2,000,000).

The new firm's earnings per share are \$4.20 (=10,500,000/2,500,000). However, the earnings increase should not favorably impact B's old shareholders. The earnings increase came as the result of a merger with a firm which the market was valuing less highly--giving a lower price/earnings ratio. Assuming that the merger offers no synergies, we would expect the new firm's price/earnings ratio to be lower than firm B's.

B13. a. The multiple of earnings paid is calculated with Equation (28.2), but with aggregate rather than per share values. The multiple of earnings paid for Firm 1 is 15 (= 450/30).

The multiple of EBIT paid is calculated using Equation (28.4) and for Firm 1 is 8.4 (= [225+450]/80).

The multiple of EBITDA paid is calculated using Equation (28.5) and for Firm 1 is 6.1 (= [225+450]/110).

The values for the other firms appear in the table below.

Firm	Multiple of Earnings Paid	Multiple of EBIT Paid	Multiple of EBITDA Paid
1	15.0	8.4	6.1
2	12.0	7.5	6.0
3	14.0	6.2	6.2
4	10.7	8.0	6.2
5	11.0	8.0	6.3
6	10.0	8.3	5.9

b. Differences in interest expense affect earnings but not EBIT or EBITDA. Since AFC does not have any debt, multiples of EBIT or EBITDA are more useful than multiples of earnings.

c.

Multiple	Range of Multiples	Value for AFC	Implied Purchase Price
Earnings	10-15×	\$50 million	\$500-750 million
EBIT	6.2-8.4×	75	465-630
EBITDA	5.9-6.3×	100	590-630

Comparative analysis may fail for many reasons. Often the failure is the result of the difficulty in identifying a well-defined group of comparable firms. For the 6 firms used as comparables in this problem, Firms 1 and 3 appear to be "different" from the others. Not only do they have the two highest multiples of earnings, but Firm 3 apparently has no D and A charges. After further investigation, an analyst might choose to eliminate these two firms from the analysis. The result would be more consistent value ranges across the three ratios.

B14. a. The market value of the $100 million of outstanding bonds is:

$$PV = \sum_{t=1}^{10} \frac{10}{(1.13)^t} + \frac{100}{(1.13)^{10}} = \$83.72 \text{ million}$$

b. Because Empire State assumes Garden States's outstanding debt, the value of the $100 million of Garden State's bonds increases to:

$$PV = \sum_{t=1}^{10} \frac{10}{(1.12)^t} + \frac{100}{(1.12)^{10}} = \$88.70 \text{ million}$$

The acquisition therefore benefits Garden State's bondholders.

c. A total of $4.98 (= 88.70 - 83.72) of wealth is transferred from Empire State's shareholders to Garden State's bondholders because of a reduction in the risk of the bonds--Empire State financially guarantees repayment of the Garden States debt when it assumes the debt repayment obligation.

B15. a. With a 1/3 premium, Big Sky will pay $32 per share for the stock in Far West. The total purchase price is $64 million (= 2×32). This $64 million purchase price plus the $500,000 of after-tax acquisition expenses gives a net acquisition cost of $64.5 million. Because Far West has no debt outstanding, the PV of debt displaced is zero and, by Equation (28.12), added debt is:

$$\text{Added Debt} = 64.5\text{ M}\times(2/3) - 0 = \$43 \text{ million}$$

The after-tax interest cost will be $3.612 million (= (1 - 0.4)×(0.14)×43 million). Deducting this from the year 8 net operating profit of $45 million gives earnings of $41.388 million. Using the current price-earnings multiple of 6, the terminal value of the equity is $248.3 million (= 6×[41.388 million]). From Equation (28.13), the terminal value of net assets is:

$$TV(NA) = TV(E) + TV(D) - T - E = 248.3 + 43 - 0 - 0 = \$291.3 \text{ million}$$

The weighted average cost of capital for the acquisition is:

$$WACC = L\times r_e + (1 - T)\times L\times r_d = (1/3)\times 20 + (1 - 0.4)\times(2/3)\times 14 = 12.27\%$$

Finally, the net present value on a non-disposition basis is:

$$NPV = -64.5 + \frac{6}{(1.1227)} + \frac{10}{(1.1227)^2} + \ldots + \frac{321.3}{(1.1227)^8} = \$125.59 \text{ million}$$

The internal rate of return is IRR = 32.91%.

b. At a purchase price of $32 per share, the acquiring multiple is 8 (= 32/4), and the terminal value on a disposition basis is $331.1 million (= 41.388×8). The long-term capital gain on the disposition will be $266.6 million (= 331.1 - 64.5), and the tax on that gain is $106.6 million (= 0.4×[266.6 million]). The resulting terminal value of net assets is:

$$TV(NA) = TV(E) + TV(D) - T - E = 331.1 + 43 - 106.6 - 0 = \$267.5 \text{ million}$$

The net present value on a disposition basis is:

$$NPV = -64.5 + \frac{6}{(1.1227)} + \frac{10}{(1.1227)^2} + \ldots + \frac{297.5}{(1.1227)^8} = \$116.13 \text{ million}$$

The internal rate of return is IRR = 32.04%.

c. Yes. Big Sky should make the acquisition if it has to pay a 1/3 premium for the shares of Far West.

B16. First, solve Equation (16.19) for T^* :

$$WACC = r - T^* \times L \times r_d \times \left[\frac{1 + r}{1 + r_d}\right]$$

$$0.1227 = 0.16 - T^* \times \left[\frac{2}{3}\right] \times (0.14) \times \left[\frac{1.16}{1.14}\right]$$

$$T^* = 0.39275$$

The annual before-tax interest cost consists of the interest on the added debt, which amounts to $6.02 million (= [0.14]×43 million). From Equation (28.15), the net APV is:

$$\text{Net APV} = -64.5 + \frac{6}{(1.16)} + \frac{10}{(1.16)^2} + \frac{12}{(1.16)^3} + \frac{14}{(1.16)^4} + \frac{18}{(1.16)^5} + \frac{22}{(1.16)^6}$$

$$+ \frac{25}{(1.16)^7} + \frac{321.3}{(1.16)^8} + \sum_{t=1}^{8} \frac{0.39275 \times (6.02)}{(1.14)^t} = \$98.94 \text{ million}$$

B17. a. With no premium, Ajax will exchange a fraction of a share worth $20 for each share of Central, and the exchange ratio will be 0.5 (= 20/40).

b. With a 25% premium, Ajax will exchange a fraction of a share worth $25 (= 20×[1.25]) for each share of Central, and the exchange ratio will be 0.625 (= 25/40).

c. With the exchange ratio fixed at 0.625, a 10% drop in the price of Ajax common stock (to $36) causes the price to be paid for each Central share to fall to $22.50 (= 36×[0.625]). As a result of this decrease in value, Central's share price would only increase to $22.50, which represents a 12.5% increase from the initial share price of $20 (i.e., a true premium of 12.5%, not 25%).

B18. Focusing on just the car rental acquisitions, the price/earnings ratios range from 10x to 12x, implying an aggregate price of $250 to 300 million. The price/cash flow ratios range from 5.8x to 7.8x, implying an aggregate price of $232 to 312 million. The price/book value ratios range from 1.4x to 1.6x, implying an aggregate price of $245 to 280 million. The (debt+equity)/EBITDA ratios range from 7.9x to 8.9x, implying an aggregate price of $237 to 267 million. The intersection of these four price ranges gives a reasonable range of acquisition values: $250 to $267 million.

B19. The adjusted present value technique is the correct one to use because of the planned decrease in leverage over the life of the investment. We determine the time pattern of tax shields and apply Equation (25.11). Because the $100 million purchase price represents an incremental free cash flow multiple of 8, the current incremental free cash flow is $12.5 million, and Spring Lake will incur $90 million of 10% debt. The after-tax interest rate is 6% (= [1 - 0.4]×10).

To calculate the Tax Shield, we need a value for T^*. This can be estimated by computing the firm's WACC and then solving Equation (16.9) for T^*.

$$\text{WACC} = 0.50\times(0.14) + 0.50\times(0.10)\times(1 - 0.40) = 10\%$$

Solving Equation (16.9) For T^*:

$$0.10 = 0.14 - T^*\times(0.50)\times(0.10)\times\frac{1.14}{1.10}$$

$$T^* = 0.7719$$

The relevant cash flows are summarized below:

Year	1	2	3	4	5	6
IFCF	13.75	15.13	16.64	18.30	20.13	22.14
After-Tax Interest	5.40	4.90	4.29	3.54	2.66	1.61
Principal Repayment	8.35	10.23	12.35	14.76	17.47	20.53
Principal Balance	81.65	71.42	59.07	44.32	26.84	6.31
Tax Shield (T^*INT_t)	4.17	3.78	3.31	2.74	2.05	1.24

The terminal value of the equity will be $177.16 million (= 22.145×[8]). The tax on the capital gain upon disposition will be $30.864 million (= 0.4×[177.16 - 100]). With $6.31 million of debt outstanding at the end of year 6, the terminal value of net assets will be $152.61 million (= 177.16 + 6.31 - 30.864).

Applying Equation (28.15), the adjusted present value of the leveraged buyout is:

$$\text{APV} = -100 + \frac{13.75}{1.14} + \frac{15.125}{(1.14)^2} + \frac{16.638}{(1.14)^3} + \frac{18.301}{(1.14)^4} + \frac{20.131}{(1.14)^5} + \frac{22.145 + 152.61}{(1.14)^6}$$

$$\frac{4.17}{1.10} + \frac{3.78}{(1.10)^2} + \frac{3.31}{(1.10)^3} + \frac{2.74}{(1.10)^4} + \frac{2.05}{(1.10)^5} + \frac{1.24}{(1.10)^6}$$

$$= \$49.08 \text{ million}$$

C1. a. Assuming straight-line depreciation to a zero book value, depreciation will be $1.6 million per year (= [8 - 0]/5). The present value of the tax savings from this depreciation, which would result from writing up the tax basis of the plants, is offset by the tax on the $8,000,000 gain. So the NPV is

$$NPV = \sum_{t=1}^{5} \frac{0.4(1{,}600{,}000)}{(1.12)^t} - 0.4(8{,}000{,}000) = -\$892{,}943$$

Because the NPV is negative, the acquiror would prefer a tax-free transaction.

b. The capital gain would be $3 million (= 8 - 5), which would be taxed $840,000 (= [0.28]3 million). To compute the after-tax cost of this tax, because it will be paid with after-tax dollars, we divide by 1 minus the tax rate, to get an after-tax cost of $1,166,667. By having a tax-free transaction, the sellers are only postponing this tax 2 years, because they are going to sell out in 2 more years anyway. So the incremental cost of a taxable transaction, ICTA, is the lost benefit to postponing is the time-value-of-money for 2 years:

$$ICTA = \left[\frac{0.28(8{,}000{,}000-5{,}000{,}000)}{1 - 0.28}\right]\left[1 - \frac{1}{(1.07)^2}\right] = \$147{,}655$$

The sellers would prefer a tax-free transaction unless they are can get $147,655 more.

c. Net advantage of a taxable transaction = -892,943 - 147,655 = -$1,040,598

So both the acquiror and the sellers would benefit from a tax-free transaction.

C2. a. With no synergistic effects, the maximum combined value, $V_{AB} = V_A + V_B = \$300$ million.

b. Because 1/3 of the combined value is due to Arnold's assets and 2/3 is due to Beard's assets, with perfect substitutes for each firm, the expected return on the shares of the merged firm in a perfect capital market environment will be 23.333% (= [1/3]× 20 + [2/3]× 25).

c. For every dollar invested, buy $0.33 of the A substitute and $0.67 of the B substitute. Short sell $1 of the merged stock. Your payoff from the substitute stock will be 0.33×(1.20) + 0.67×(1.25) = $1.2335. Then, pay the $1.23 required return on the merged stock for a net riskless profit of $.0035 per dollar invested.

d. If neither firm has a close substitute, the merger reduces the set of possible outcomes from investing in the two assets to the single outcome for the merged firm. This reduction in the outcome space spanned by the investment reduces the value of the combined investment from its pre-merger value.

C3. a. Because the likelihood that state F will occur for firm A and the likelihood that it will occur for firm B are uncorrelated, there are 3 possible states for the merged firm. FF is the state when both of the unmerged firms would have been in state F, which occurs with probability 0.25 (= [0.5]×0.5). FU is the state when one of the two unmerged firms would have been in state F and the other would have been in state U; there are two ways this can happen--what would have been firm A is in state F and what would have been firm B is in state U or vice versa. As a result, the probability of state FU for the merged firm is 0.5 (= 2×[0.5]×0.5). UU is the state when both of the unmerged firms would have been in state U, which occurs for the merged firm with probability 0.25. As a result of the stock-for-stock merger, the bondholders are treated as a single group and the stockholders are treated as a single group regardless of which set of assets really earned a particular return, so we have the following set of outcomes for the merged firm:

	Return in State			Expected	Standard
Return to	FF	FU	UU	Return	Deviation
Merged firm	400	300	200	300	70.71
Debtholders					
Former A	140	140	100	130	17.32
Former B	140	140	100	130	17.32
Stockholders					
Former A	60	10	0	20	23.45
Former B	60	10	0	20	23.45

b. The debtholders are unambiguously better off as a result of the merger because they have a higher expected return and a lower standard deviation.

c. The stockholders are generally worse off as a result of the merger because their expected return has fallen by 1/3 (= [30 - 20]/30) while the standard deviation of their return has fallen by only 22% (= [30 - 23.45]/30).

d. The conclusion that can be drawn from parts b and c is that the two firms are effectively guaranteeing each other's debt, and wealth is transferred from the stockholders to the bondholders as a result of the merger.

C4. a. The year 10 incremental free cash flow is \$60.683 million (= $15\times(1.15)^{10}$) so the terminal value will be \$505.69 million (= $[125/15]\times 60.683$) and tax on the capital gain will be \$152.276 million (= $0.4\times[505.69 - 125]$). The net terminal value will be \$353.414 million (= 505.69 - 152.276).

The net effect of the taxable transaction is a \$10 million depreciation recapture tax liability with additional depreciation expense of \$5 million per year for 5 years. The NPV of a taxable transaction is

$$(0.4)\times\sum_{t=1}^{5}\frac{5}{(1.10)^t} - 10 = -\$2.42 \text{ million}$$

Radnor's NPV of acquiring Excelsior is:

$$NPV = -125 + \sum_{t=1}^{10}\frac{15\times(1.15)^t}{(1.16)^t} + \frac{353.414}{(1.16)^{10}} - 2.42 = \$95.76 \text{ million}$$

b. Radnor's NPV of acquiring Excelsior's common stock in a tax-free transaction is:

$$NPV = -125 + \sum_{t=1}^{10}\frac{15\times(1.15)^t}{(1.16)^t} + \frac{353.414}{(1.16)^{10}} = \$98.18 \text{ million}$$

c. Because Miami Media would have to realize a $25 million taxable gain in a taxable transaction, it would prefer a tax-free transaction. The present value difference in taxes payable is:

$$25 \times \left[1 - \frac{1}{(1.15)^{10}}\right] = \$18.82 \text{ million}$$

d. Miami Media and Radnor are both better off, from a purely tax standpoint, in a tax-free transaction. Unless there are nontax reasons for structuring the acquisition in a taxable manner, they should structure it as a tax-free acquisition.

C5. a. Time's board of directors, which included members of Time's management, refused to let the shareholders vote on the Warner Communications transaction because they feared that many--perhaps a majority--would vote against it. Clearly, the Paramount bid had substantially greater *immediate* value to Time's shareholders than the Time-Warner merger, but the management of Time had already agreed with Warner's management as to the management structure of what would become Time-Warner. In any case, Time's board of directors refused to allow Time's shareholders to choose between the two transactions.

b. Time's board of directors may have possessed material nonpublic information, which it felt it could not disclose for competitive or other reasons, that suggested that after the Time-Warner merger the new firm would be worth more than $200 per original Time share. If so, then asymmetric information would account for the difference between the current share price of $144 per share and the present value under the firm's long-term strategy being in excess of $200 per share.

C6. a. In early April 1996, the Brooke Group conceded that management had won.

b. From the initiation of the proxy contest until its resolution, RJR Nabisco stock fell about 3% in value. Over the same time period, the S&P 500 rose by about 3%.

CHAPTER 29 -- MULTINATIONAL CORPORATE FINANCE

A1. Forward contracts specify the purchase and sale of a currency for future delivery based on exchange rates that are agreed to today and generally have a term of between 1 and 52 weeks. The advantage of forward contracts is that they can be customized to suit a corporation's particular requirements as to amount and settlement date. A futures contract is a standardized forward contract that is exchange-traded. The advantage of futures contracts is that they are generally less costly and are more liquid than forward contracts; a firm can close its position in a futures contract at any time simply by selling (or repurchasing) the contract.

A2. a. One Canadian dollar equals 1.2526 Swiss francs.

b. One Swiss franc equals 0.79834 Canadian dollars.

c. 1/1.2526 Swiss francs/Canadian dollar = 0.79834 Canadian dollars/Swiss franc, so the cross rates in a and b are equivalent.

A3. a. The German mark is trading at a forward discount, except for the 30-day forward rate which is at a slight premium over the spot rate.

30-day forward premium = $12 \times [(1.6925 - 1.6924)/1.6925] \times 100$ = 0.07% premium

60-day forward discount = $6 \times [(1.6937 - 1.6925)/1.6925] \times 100$ = 0.43% discount

90-day forward discount = $4 \times [(1.6971 - 1.6925)/1.6925] \times 100$ = 1.09% discount

b. The Japanese yen is trading at a forward premium to the U. S. dollar.

30-day forward premium = $12 \times [(144.70 - 144.53)/144.70] \times 100$ = 1.41% premium

60-day forward premium = $6 \times [(144.70 - 144.33)/144.70] \times 100$ = 1.53% premium

90-day forward premium = $4 \times [(144.70 - 144.04)/144.70] \times 100$ = 1.82% premium

A4. a. Interest rate parity requires the forward-spot exchange rate differential to offset the interest rate differential between two countries.

b. Purchasing power parity states that the differential between the expected inflation rates in two countries equals the differential between the current and expected future spot exchange rates.

c. The expectations theory states that the forward exchange rate equals the expected future spot exchange rate.

d. The preceding three relationships together provide for the international Fisher effect; this relationship requires that the expected real rates of interest must be the same in all the world's capital markets.

A5. (1.125 DM/SF)/(96.18 ¥/SF) = 0.0117 DM/¥, so the Japanese yen-German mark exchange rate is ¥1 = DM 0.0117.

A6. Under purchasing power parity,

$$\frac{E[1 + I_{£}(t)]}{E[1 + I_{\$}(t)]} = \frac{s_{\$/£}}{E[s_{\$/£}(t)]}$$

$$E[s_{\$/£}] = (2.00/£)\times(1.04/1.08) = \$1.93/£$$

A7. a. If the forward rate is greater than the expected future spot rate, then everyone will want to sell pounds forward but no one will want to sell dollars forward.

b. If the forward rate is less than the expected future spot rate, then everyone will want to sell dollars forward but no one will want to sell pounds forward.

A8. Statement b is a stronger requirement because it requires both that statement a hold and that foreign exchange market participants are perfect forecasters.

A9. Equation (29.5) requires that

$$\frac{1 + r_{SF}(t)}{1 + r_{\$}(t)} = \frac{E[1 + i_{SF}(t)]}{E[1 + i_{\$}(t)]}$$

Since 1.04/1.08 = 0.096 and 1.01/1.05 = 0.96, Equation (29.5) holds. The real rate of interest is

$$\frac{1 + r_n}{1 + i} - 1 = \frac{1.08}{1.05} - 1 = \frac{1.04}{1.01} - 1 \approx 3\% \text{ per year)}$$

A10. The difference between the current spot exchange rate and the forward exchange rate is not an accurate measure of the cost of a forward contract because the true cost of a forward contract is its opportunity cost, which depends on the proceeds the firm would realize if it did not hedge; therefore, the cost depends on the difference between the forward rate and the expected future spot rate.

A11. A foreign currency money market hedge can be thought of as a "homemade" forward contract in the sense that, when interest rate parity holds, any difference between selling a foreign currency forward (as in the forward market hedge) and selling it spot (as in the foreign currency money market hedge) will be fully offset by the difference between the interest rates in the two currencies. From the standpoint of practical foreign currency risk management, managers should choose whichever method is favored by government restrictions on forward selling or on foreign-currency-denominated borrowing.

A12. If the future spot rate is £1 = \$1.25, the value of the original contract is £100 million = \$125 million but Pratt and Whitney delivers the £100 million for \$160 million under the forward contract for a gain of \$35 million. The net realized value is \$160 million (= 125 + 35). With the foreign currency money market hedge, when the spot rate falls to £1 = \$1.25, the £100 million of principal and interest will cost \$125 million to repay, while the dollar investment returns \$160 million for a \$35 million gain on the money market hedge. Combined with the \$125 million value of the original contract, the net realized value is again \$160 million.

continued . . .

A12. cont. If the future spot rate is £1 = \$2.00, the value of the original contract is £100 million = \$200 million but Pratt and Whitney delivers the £100 million for \$160 million under the forward contract for a loss of \$40 million. The net realized value is \$160 million (= 200 - 40). With the foreign currency money market hedge, when the spot rate rises to £1 = \$2.00, the £100 million of principal and interest will cost \$200 million to repay, while the dollar investment returns \$160 million for a \$40 million loss on the money market hedge. Combined with the \$200 million value of the original contract, the net realized value is again \$160 million.

A13. The U.S. dollar internal rate of return is the discount rate that solves the following equation:

$$0 = -30 + \frac{9.17}{(1 + IRR)} + \frac{10.59}{(1 + IRR)^2} + \frac{12.69}{(1 + IRR)^3} + \frac{12.93}{(1 + IRR)^4} + \frac{11.20}{(1 + IRR)^5}$$

IRR = 24.35%. Because the IRR is greater than the 15% required return, General Electric should undertake the project.

The Deutsche mark internal rate of return is the discount rate that solves the following equation:

$$0 = -DM50 + \frac{15}{(1 + IRR)} + \frac{17}{(1 + IRR)^2} + \frac{20}{(1 + IRR)^3} + \frac{20}{(1 + IRR)^4} + \frac{17}{(1 + IRR)^5}$$

IRR = 22.04%. Because the IRR is greater than the 12.87% required return, General Electric should undertake the project.

Recall from Chapter 11 that the NPV criterion is the only one of the two that is universally valid, although the IRR and NPV criteria can be used interchangeably for economically independent capital investment projects when there is just one change in sign in the net cash flow.

A14. If interest rate parity holds, a currency with a relatively low interest rate has that low rate because market participants expect the country to have a relatively low inflation rate and expect its currency to appreciate in value. Nominal interest cost savings will be offset exactly by expected exchange rate changes. The wise corporate treasurer will evaluate international capital market frictions that may make it advantageous to borrow in a particular currency rather than follow a policy of always borrowing in whichever currency affords the lowest interest rate.

B1. a. The after-tax interest payment on the domestic debt is \$25 (= (1 - 0.5)× 1000 ×[0.10/2]) per semi-annual period, and the net proceeds are \$990 (= (1 - 0.01)× 1000) per bond. Amortization of the issuance expenses provides a tax shield of 0.5 ×(1000)×(0.01)/12 = 0.417 per period so that the semiannual cash flow is 24.583 (= 25 - 0.417). Thus, the after-tax cost of the domestic issue is the discount rate, *r*, that solves the following equation:

$$990 = \sum_{t=1}^{12} \frac{24.583}{(1 + r)^t} + \frac{1000}{(1 + r)^{12}}$$

r = 2.56% per semiannual period, an APY of 5.19%.

b. The after-tax interest payment on the Eurobonds is \$51.25 (= (1 - .5)×1000×[0.1025]) per year, and the net proceeds are \$987.50 (= (1 - 0.0125)× 1000) per bond. Amortization of the issuance expenses provides a tax shield of 0.5×(1000)×(0.0125)/6 = 1.042 per period so that the annual cash flow is 50.208 (= 51.25 - 1.042). Thus, the after-tax cost of the Eurobond issue is the discount rate, r, that solves the following equation:

$$987.50 = \sum_{t=1}^{6} \frac{50.208}{(1 + r)^t} + \frac{1000}{(1 + r)^6}$$

r = 5.27% per year.

c. The domestic issue is cheaper, after-tax it has an APY of 5.19% versus an APY of 5.27% for the Eurobond issue.

B2. a. The after-tax cost of the U.S.-dollar-denominated issue is 6.60% (= (1 - 0.34)×10).

b. The cash flows associated with the Deutsche-mark-denominated issue are summarized in the table below:

Year	1	2	3	4	5
Principal (DM)	-	10.00	10.000	15.000	15.000
Interest (DM)	4.00	4.00	3.200	2.400	1.200
Tax saving (34%)	-1.36	-1.36	-1.088	-0.816	-0.408
Total	2.64	12.64	12.112	16.584	15.792
Exchange rate	0.6113	0.6228	0.6345	0.6465	0.6586
Debt Service (\$)	1.6138	7.8722	7.6851	10.7216	10.4006

The after-tax cost of the Deutsche-mark-denominated issue expressed in U.S. dollars is the discount rate, IRR, that solves the following equation:

$$0 = -30.0 + \frac{1.6138}{(1 + \text{IRR})} + \frac{7.8722}{(1 + \text{IRR})^2} + \frac{7.6851}{(1 + \text{IRR})^3} + \frac{10.7216}{(1 + \text{IRR})^4} + \frac{10.4006}{(1 + \text{IRR})^5}$$

so the after-tax cost is IRR = 7.26%.

B3. a.

$$\frac{1 + r_{SF}(t)}{1 + r_{\$}(t)} = \frac{s_{\$/SF}}{f_{\$/SF}(t)}$$

Therefore, 1.03/1.08 = $0.66/f_{\$/SF}(t)$; and the 3-month forward rate is SF1 = \$0.6921

b. Borrow SF1515.15, buy \$1000 (= SF1515.15×[0.66\$/SF]), and sell \$1080 forward for 3-month delivery. Invest the dollars to yield \$1080 (= 1000 + [0.08]×1000). Deliver the \$1080 under the forward contract for SF1588.24 (= 1080/0.68).

c. Use SF1560.60 (= 1515.15 + [0.03]×1515.15) to repay the loan, which leaves a riskless arbitrage profit of SF27.64 (= 1588.24 - 1560.60) or \$18.79 (= [0.68]×27.64) for every \$1000 invested.

B4. a.

$$\frac{1 + r_{\$}(t)}{1 + r_{DM}(t)} = \frac{s_{DM/\$}}{f_{DM/\$}(t)}$$

Therefore, $(1 + r_{\$}(t))/1.05 = 1.50/1.40$; and the 1-year U.S. dollar interest rate is 12.5%.

b. $1.10/(1 + r_{DM}(t)) = 1.50/1.40$; so the 1-year German mark interest rate is 2.67%.

c. For interest rate parity to hold,

$$\frac{1 + r_{\$}(t)}{1 + r_{DM}(t)} = \frac{1.50}{1.40} = 1.07143$$

B5. a. For the Law of One Price to hold, \$450 = £300 so the dollar-pound exchange rate would have to be \$1.50/£.

b. Buy 1 oz. of gold in New York for \$450. Transport it to London and sell it for £300. Exchange the pounds for \$510.

c. This riskless arbitrage provides a profit of \$60, or £35.29, for every ounce of gold transported.

B6. If the expectations theory of forward exchange rates holds, then $E[s_{D/F}(t)] = f_{D/F}(t)$ and this relationship can be substituted into the purchasing power parity relationship to give:

$$\frac{E[1+i_F(t)]}{E[1+i_D(t)]} = \frac{s_{D/F}}{f_{D/F}(t)}$$

If interest rate parity also holds, then

$$\frac{1+r_F(t)}{1+r_D(t)} = \frac{s_{D/F}}{f_{D/F}(t)} = \frac{E[1+i_F(t)]}{E[1+i_D(t)]}$$

As a result, the international Fisher effect must hold when interest rate parity, purchasing power parity, and the expectations theory of forward exchange rates all hold. If any one of the first three parity relationships fails to hold, then the international Fisher effect does not necessarily hold, and we would expect to see different real interest rates across different currencies.

B7. a. The best estimate of the spot rate expected 1 year from now is the 1-year forward rate, SF1 = DM1.30.

b. If the expected spot rate was SF1 = DM1.40, everyone would want to sell Swiss francs forward but no one would be willing to sell German marks forward.

c. If the expected spot rate was SF1 = DM1, everyone would want to sell German marks forward but no one would be willing to sell Swiss francs forward.

B8. Under purchasing power parity, we have:

$$\frac{E[1+i_Y(1)]}{E[1+i_{DM}(1)]} = \frac{s_{DM/Y}}{E[s_{DM/Y}(1)]} = \frac{0.012}{0.015} = 0.8 \text{ so that } \frac{E[1+i_{DM}(1)]}{E[1+i_Y(1)]} = \frac{1}{0.8} = 1.25$$

Thus, the inflation rate in Germany for the coming year is expected to be 25 percentage points greater than the expected inflation rate in Japan (that is, 30% versus 5%).

B9. According to the international Fisher effect, we have:

$$\frac{1+r_C(5)}{1+r_{GB}(5)} = \frac{E[1+i_C(5)]}{E[1+i_{GB}(5)]} = \frac{1.07}{1.10} = 0.9727$$

Thus, the ratio of 5-year interest rates in Canada to 5-year interest rates in Great Britain would be 0.9727. The 5-year interest rate in Great Britain should exceed the 5-year interest rate in Canada by approximately 3% per year.

B10. a. The expected future spot rate is equal to the forward exchange rate so TransAtlantic would expect to receive $2.9 million (= DM5 million ×[$0.58/DM]) if it does not hedge.

b. With hedging, TransAtlantic could have locked in the expected $2.9 million payment. If the spot rate at the time the tour operator pays is DM1 = $0.56, then TransAtlantic will actually receive $2.8 million (= DM 5 million ×[$0.56]) for a loss of $100,000.

c. Borrow DM4,545,455 (= DM5 million/1.10) for 30 days. Convert it into $2,727,273 (= DM4,545,455 ×[$0.60/DM1]) and invest the dollar proceeds at 10% for 30 days to realize $3 million. This exceeds the $2.9 million that would be locked in with a forward contract so the foreign currency money market transaction is the preferred hedging method in this situation.

d. A foreign currency money market transaction as in part c. would produce $2,894,737, which is less than the $2.9 million that would be locked in with a forward contract, so the forward contract is the preferred hedging method in this situation.

B11. With the forward contract, TransAtlantic can insure a payoff of $2.9 million. To receive the same payoff from a dollar investment at 12% interest, the firm would have to invest $2,589,286 (= 2.9 million/1.12). With a spot exchange rate of DM1 = $0.60, this is equivalent to DM4,315,477 (= 2,389,286/0.60). For the firm to be indifferent between the forward contract and the foreign currency money market transaction, the DM investment must yield DM5 million. This requires a German 30-day interest rate of 15.862% (= (5,000,000/4,315,477) - 1).

B12. From the equation describing nominal interest rates, we have the following relationship:

$(1 + r_r) = (1 + r_n)/(1 + i)$

Given that the 1-year riskless rate is 8% in the U.S. and the expected inflation rate in the U.S. is 5% per year for the next 5 years, we can compute the expected real rate of interest in the U.S: $r_r = (1.08)/(1.05) - 1 = 2.86\%$.

The 1-year riskless rate in Germany is 6% and the expected inflation rate in Germany is 3.06%, so the same calculation leads to the expected real rate of interest in Germany: $r_r = (1.06)/(1.0306) - 1 = 2.85\%$.

When interest rate parity, purchasing power parity, and the expectations theory of forward exchange rates all hold, then the international Fisher effect holds and the ratio of nominal interest rates is equal to the ratio of expected inflation rates. By cross-multiplying that relationship, the ratio of the nominal interest rate to the expected inflation rate is equal across countries. If this is true, then, by the initial equation given above, the real interest rates must be equal across countries.

B13. a. To find the expected spot exchange rates, we can use the combined market equilibrium condition in Figure 29-3:

$$\frac{1 + r_{DM}(t)}{1 + r_{\$}(t)} = \frac{s_{\$/DM}}{E[s_{\$/DM}(t)]}$$

and solve for $E[s_{\$/DM}] = s_{\$/DM} \times (1 + r_{\$})/(1 + r_{DM})$. When the interest rate is expressed as a rate per year, the expected spot rate satisfies the equation:

$$E[s_{\$/DM}(t)] = s_{\$/DM} \times [(1 + r_{\$}(t))/(1 + r_{DM}(t))]^t$$

With the current spot rate of DM1 = \$0.60 and the interest rates provided, we can compute the expected spot rate for each year:

Year	1	2	3	4	5
$E[s_{\$/DM}(t)]$	0.6113	0.6226	0.6168	0.6223	0.6000

b. Multiplying each year's cash flow in German marks by the expected spot rate gives the expected incremental cash flow stream in dollars:

Year	1	2	3	4	5
CF(\$M)	9.1695	10.5842	12.336	12.446	10.200

c. The NPV of the project is:

$$NPV = -30 + \frac{9.1695}{(1.15)} + \frac{10.5842}{(1.15)^2} + \frac{12.336}{(1.15)^3} + \frac{12.446}{(1.15)^4} + \frac{10.20}{(1.15)^5} = \$6{,}275{,}018$$

B14. a. Under interest rate parity and the expectations theory of forward exchange rates, the expected spot rate 1 year hence satisfies:

$$\frac{1.12}{1.08} = \frac{1.50}{E[s_{\$/£}(1)]}$$

so that $E[s_{\$/£}(1)] = \$1.4464$

Under purchasing power parity,

$$\frac{E[1 + I_£(1)]}{1.05} = \frac{1.50}{1.4464}$$

$$E[1 + i_£(1)] = 1.0889$$

Thus, the expected inflation rate in the United Kingdom is 8.89%.

b. The expected spot rate t years from now satisfies:

$$E[s_{\$/£}(t)] = 1.50 \times (1.05/1.0889)^t$$

The expected spot exchange rates ($/£) will be:

Year	Exchange rate
1	1.4464
2	1.3947
3	1.3449
4	1.2969
5	1.2505
6	1.2059
7	1.1628

c. The incremental cash flow stream in dollars is computed by applying the appropriate expected spot rate to the cash flows in pounds:

Year	Cash flow ($ millions)
1	43.392
2	55.788
3	53.796
4	64.845
5	62.525
6	60.295
7	58.140

d. The NPV in dollars is computed by discounting the cash flows in part c at the dollar hurdle rate of 14% and deducting the initial cost of $100 million. This calculation gives NPV = $138.9 million.

e. We estimate the pound sterling hurdle rate by applying the relationship in Equation (29.7):

$(1.12)/(1.08) = (1 + r^*_£)/(1.14)$; so that $r^*_£ = 18.22\%$

f. The NPV in pounds is computed by discounting the cash flows in pounds at the pound sterling hurdle rate from part e and deducting the initial cost of £66.67 million. This method gives NPV = £92.6 million.

g. The NPV in part f is £92.6 million. Converting this amount to dollars gives £92.6 million ($1.50/£) = $138.9 million. This is equal to the NPV calculated in dollars in part d.

B15. The debt service cash flows for the General Electric example are presented in the following table:

Year	1	2	3	4	5
Principal (DM)	-	10	10	15	15
Interest (DM)	4	4	3.2	2.4	1.2
Tax saving on interest	-2	-2	-1.6	-1.2	-0.6
Total	2	12	11.6	16.2	15.6
Exchange rate	0.6113	0.6228	0.6345	0.6465	0.6586
Debt Service ($)	1.2226	7.4736	7.3602	10.4733	10.2742
Tax Savings on currency appreciation*	-0.0113	-0.1368	-0.2001	-0.37665	-.45708
Adjusted debt service ($)	1.2113	7.3368	7.1601	10.09665	9.81712

*(ER - 0.6000)×(Total debt service payment)×(0.5)

For example, in year 1, the tax savings on currency appreciation are equal to:

$(0.6113 - 0.6000)\times(2)\times(0.5) = 0.0113$

The after-tax cost of the Deutsche-mark-denominated debt is the discount rate, IRR, that solves the following equation:

$$0 = -30 + \frac{1.2113}{(1 + IRR)} + \frac{7.3368}{(1 + IRR)^2} + \frac{7.1601}{(1 + IRR)^3} + \frac{10.09665}{(1 + IRR)^4} + \frac{9.81712}{(1 + IRR)^5}$$

So the after-tax cost is IRR = 4.992%

C1. The APY for the 2-year interest rate of 10% per year compounded semiannually is 10.25% (= $[1.05]^2$ - 1). The APY for the 1-year interest rate of 8% per year compounded semiannually is 8.16% (= $[1.04]^2$ - 1). The APY for the 1-year forward rate is the rate f that solves the following equation:

$(1.1025)^2 = (1.0816)\times(1 + f)$; so that $f = 12.38\%$

C2. a. With continuous compounding, the interest rate parity becomes

and interest rate parity takes the form

$$e^{r_£(t) - r_\$(t)} = s_{\$/£}/f_{\$/£}(t) \tag{29.2'}$$

b. With continuously compounded inflation rates, purchasing power parity becomes

$$\frac{E[e^{i_£(t)}]}{E[e^{i_\$(t)}]} = \frac{s_{\$/£}}{E[s_{\$/£}(t)]} \tag{29.3'}$$

c. Substituting Equation (29.4), $E[s_{\$/£}(t)] = f_{\$/£}(t)$, into Equation (29.3'), we have:

$$\frac{E[e^{i_£(t)}]}{E[e^{i_\$(t)}]} = \frac{s_{\$/£}}{f_{\$/£}(t)}$$

Then further substituting Equation (29.2') for the right hand side, we have

$$\frac{E[e^{i_£(t)}]}{E[e^{i_\$(t)}]} = e^{r_£(t) - r_\$(t)} \tag{29.5'}$$

d. If $r_£(1) = 0.12$, $r_\$(1) = 0.10$, $i_£(1) = 0.10$, and $i_\$(1) = 0.08$, then

$$\frac{E[e^{i_£(1)}]}{E[e^{i_\$(1)}]} = \frac{e^{0.10}}{e^{0.08}} = e^{0.02} = e^{0.12-0.10} = \frac{e^{0.12}}{e^{0.10}} = e^{r_£(1) - r_\$(1)}$$

so that our derived (modified) Equation (23.5') holds exactly.

e. The real rate of interest is 2% in both markets when expressed on a continuously compounded basis. Let r_r denote the real rate of interest, r_n the nominal rate of interest, and i the expected inflation rate, all expressed on a continuously compounded basis. Then

$$e^{r_n} = (e^{r_r}) \times (e^{i}) = e^{r_r + i} \text{ so that } r_n = r_r + i \text{ and } r_r = r_n - i$$

For the United Kingdom (£),

$r_r = 0.12 - 0.10 = 2\%$ per year

and for the United States ($),

$r_r = 0.10 - 0.08 = 2\%$ per year

C3. Suppose the expectations theory of forward exchange rates does not hold. Then $E[s_{D/F}(t)] \neq f_{D/F}(t)$. This implies that $s_{D/F}/f_{D/F}(t) \neq s_{D/F}/E[s_{D/F}(t)]$. Then if interest rate parity and purchasing power parity do hold, the international Fisher effect cannot hold.

C4. Suppose a manufacturer expects to receive an amount A in pounds sterling t periods in the future. The manufacturer could sell £A forward, deliver the £A against the forward contract, and receive an amount $[f_{\$/£}(t)]\times[£A]$ in U.S. dollars.

Alternatively, the manufacturer could borrow $£A/(1 + r_£(t))$ in pounds sterling; convert that amount into U.S. dollars at the prevailing spot rate to obtain $s_{\$/£}\times[£A/(1 + r_£(t))]$; and invest this amount for t periods at interest rate $r_\$(t)$. At the end of t periods the manufacturer would realize

$$[1 + r_\$(t)]\times[s_{\$/£}]\frac{£A}{1 + r_£(t)} = s_{\$/£} \times \frac{1 + r_\$(t)}{1 + r_£(t)} \times £A$$

The manufacturer would use the £A to repay the sterling loan ($= £A/[1 + r_£(t)]$) plus interest ($= [£A\times r_£(t)]/[1 + r_£(t)]$).

Under interest rate parity,

$$\frac{1 + r_£(t)}{1 + r_\$(t)} = \frac{s_{\$/£}}{f_{\$/£}(t)} \iff f_{\$/£}(t) = s_{\$/£} \times \frac{1 + r_\$(t)}{1 + r_£(t)}$$

so that the money market hedge yields

$$s_{\$/£} \times \frac{1 + r_\$(t)}{1 + r_£(t)} \times £A = [f_{\$/£}(t)]\times £A$$

which is an amount equal to what the forward contract hedge provides.

C5. a. Total payoff = Forward contract payoff + Sale Proceeds

$= (\$160 \text{ million} - (£100 \text{ million})\times s_{\$/£}(1)) + (£100 \text{ million})\times s_{\$/£}(1) = \$160 \text{ million}$

b. Total payoff = $[£100 \text{ million}/(1 + r_£)]\times(s_{\$/£})\times(1 + r_\$)$; rearranging,

Total payoff = $£100 \text{ million} \times[(s_{\$/£})\times(1 + r_\$)/(1 + r_£)]$

If interest rate parity holds,

$[(s_{\$/£})\times(1 + r_\$)/(1 + r_£)] = f_{\$/£}(1)$

Accordingly, total payoff = $(£100 \text{ million})\times f_{\$/£}(1) = \$160$ million.

C6. Let S_t denote the price of the foreign security at time t denominated in the local currency and let D_t denote the dividend payment at time t also denominated in the local currency. By definition,

$$({}^{F}R_t) = \frac{S_t - S_{t-1} + D_t}{S_{t-1}}$$

so that

$$1 + ({}^{F}R_t) = \frac{S_t + D_t}{S_{t-1}}$$

The total return in U.S. dollars is calculated by converting S_{t-1}, S_t, and D_t into U.S. dollars. Assuming the expectations theory of forward exchange rates holds:

$$1 + ({}^{\$}R_t) = \frac{f_{\$/F}(t) \times S_t + f_{\$/F}(t) \times D_t}{s_{\$/F} \times S_{t-1}}$$

$$= \frac{f_{\$/F}(t)}{s_{\$/F}} \times \left[\frac{S_t + D_t}{S_{t-1}}\right] = \frac{f_{\$/F}(t)}{s_{\$/F}} \times \left[1 + ({}^{F}R_t)\right]$$

Thus,

$$\frac{s_{\$/F}}{f_{\$/F}(t)} = \frac{1 + ({}^{F}R_t)}{1 + ({}^{\$}R_t)}$$

It follows from purchasing power parity and the expectations theory of forward exchange rates that

$$\frac{s_{\$/F}}{f_{\$/F}(t)} = \frac{E[1 + i_F(t)]}{E[1 + i_{\$}(t)]} = \frac{1 + ({}^{F}R_t)}{1 + ({}^{\$}R_t)}$$

Finally, what's true for each security must be true for the market as a whole, so that

$$\frac{s_{\$/F}}{f_{\$/F}(t)} = \frac{E[1 + i_F(t)]}{E[1 + i_{\$}(t)]} = \frac{1 + ({}^{F}R_{mt})}{1 + ({}^{\$}R_{mt})}$$

C7. Let β_F denote the value of beta when calculated in the foreign currency and $\beta_\$$ denote the value when calculated in U.S. dollars. By definition,

$$\beta_F = \frac{cov({}^F R_t, {}^F R_{mt})}{var({}^F R_{mt})}$$

Adding the constant 1 to each variable will not alter either the covariance or the variance so that

$$\beta_F = \frac{cov(1 + ({}^F R_t), 1 + ({}^F R_{mt}))}{var(1 + ({}^F R_{mt}))}$$

Let

$$\alpha = \frac{1 + ({}^\$ R_t)}{1 + ({}^F R_t)} = \frac{1 + ({}^\$ R_{mt})}{1 + ({}^F R_{mt})}$$

Then

$$\beta_\$ = \frac{cov(1 + ({}^\$ R_t), 1 + ({}^\$ R_{mt}))}{var(1 + ({}^\$ R_{mt}))}$$

$$= \frac{cov[\alpha \times (1 + ({}^F R_t)), \alpha \times (1 + ({}^F R_t))]}{var[\alpha \times (1 + ({}^F R_{mt}))]}$$

$$= \frac{\alpha^2 \times cov(1 + ({}^F R_t), 1 + ({}^F R_{mt}))}{\alpha^2 \times var(1 + ({}^F R_{mt}))}$$

$$= \beta_F$$

C8. By definition,

$$r^*_F = {}^F r_f(t) + \beta_F \times [{}^F R_{mt} - {}^F r_f(t)]$$

where the subscript f denotes the riskless rate in the indicated currency, and

$$r^*_\$ = {}^\$ r_f(t) + \beta_\$ \times [{}^\$ R_{mt} - {}^\$ r_f(t)]$$

The first equation can be rewritten as

$$1 + r^*_F = 1 + {}^F r_f(t) + \beta_F \times [(1 + {}^F R_{mt}) - (1 + {}^F r_f(t))]$$

From the parity relationships and from the equilibrium condition established in problem C6:

$$\frac{1 + ({}^F r_f(t))}{1 + ({}^\$ r_f(t))} = \frac{E[1 + i_F(t)]}{E[1 + i_\$(t)]} = \frac{1 + ({}^F R_{mt})}{1 + ({}^\$ R_{mt})}$$

Denote the common value of these ratios as α. Problem C7 established that $\beta_F = \beta_\$$. Then

$$1 + r^*_F = \alpha \times [1 + {}^\$ r_f(t)] + \beta_\$ \times [\alpha(1 + {}^\$ R_{mt}) - \alpha \times (1 + {}^\$ r_f(t))]$$

$$= \alpha \times [1 + {}^\$ r_f(t) + \beta_\$ \times ({}^\$ R_{mt} - {}^\$ r_f(t))]$$

$$= \alpha \times [1 + r^*_\$]$$

so that

$$\alpha = \frac{1 + r_F^*}{1 + r_\$^*} = \frac{1 + ({}^F r_f(t))}{1 + ({}^\$ r_f(t))}$$

C9. a. Dividing the price of a Big Mac in each country by the price in the United States gives the implied exchange rate for each country:

Country	Implied Exchange Rate
Australia	A$1.05/US$
Britain	£0.64/US$
Canada	C$0.995/US$
France	FF8.05/US$
Japan	¥168.18/US$
W. Germany	DM1.95/US$

b. The prices are inconsistent with the Law of One Price.

c. The observed prices do not imply that exchange rates are out of equilibrium or that the Law of One Price is invalid. It simply means that there are transaction costs or other market imperfections that prevent the Law of One Price from holding exactly. For example, in contrast to a financial asset, Big Macs are perishable and it would be impossible to transport them from one market to another without loss.